Second Edition

CHEMISTRY COUNTS

GRAHAM HILL

Hodder & Stoughton

A MEMBER OF THE HODDER HEADLINE GROUP

Contents

Preface

This book is about chemistry and the part that chemistry plays in our everyday lives, in industry and in society. I hope you will find it lively, colourful and interesting.

This second edition has been revised and updated to meet the requirements of the new syllabuses for GCSE Chemistry in Key Stage 4 of the National Curriculum.

It is divided into twelve major sections. These sections are divided into two-page units, each of which ends with several short questions to help you understand the topic you have just studied.

At the end of each section, there are Activities and longer Study Questions. The Activities are related to different aspects of chemistry and include creative writing, text-related exercises, data response questions, surveys and problem-solving exercises. They will enable you to practise different skills and help you to appreciate the wider social, environmental, economic and technological aspects of chemistry. The Study Questions are like the questions on GCSE papers. Try to answer as many of these as possible. They will help you to understand the subject and give you valuable practice in answering exam questions.

If you are looking for information on a particular topic, look it up in the contents list at the beginning of the book and in the index at the end of the book.

Safety considerations have been given the highest priority throughout the entire book and in this respect I have appreciated the experienced advice of Dr Peter Borrows, Chairman of the Safeguards in Science Committee of the Association for Science Education.

Many people have influenced the planning and writing of *Chemistry Counts* and the revision for this second edition. I am particularly grateful to my publishers, my wife Elizabeth and countless students at Dr. Challoner's.

It has been a privilege and a pleasure to continue working on a successful project with such able, enthusiastic and supportive colleagues.

Graham Hill
Dr Challoner's Grammar School,
Amersham

To Aunt Clara

SECTION A
From Raw Materials to Pure Substances

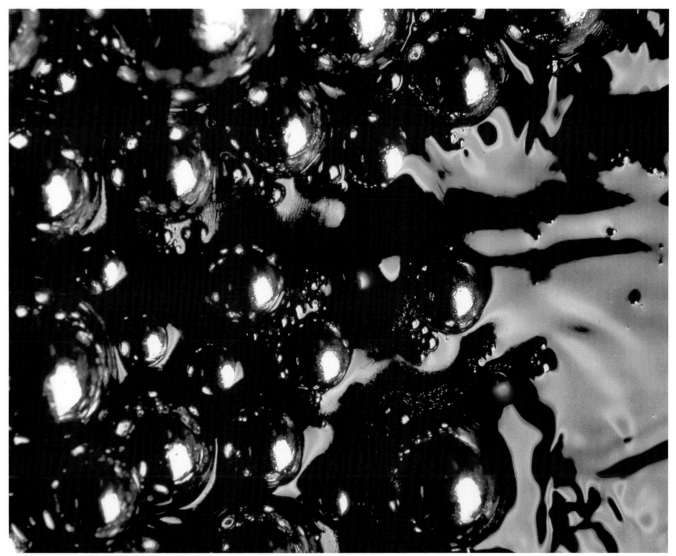

Black gold—crude oil is an important source of chemicals and fuels

1 Raw Materials

Mining iron ore

Iron ore can be made into steel and used to build structures like the Humber Bridge

Iron ore, like that in the photograph above, is useless. We can't eat it, wear it or grow things in it. But, if we heat it with limestone, coke and air we can turn it into iron, and from iron we can make steel.

These different materials like iron, iron ore, limestone, coke and air are called **substances**.

> * The study of materials and substances is called **chemistry**.
> * Chemists and chemical engineers study materials and try to change useless materials like iron ore into useful substances like iron and steel.

Raw materials from the Earth

Materials like iron ore, limestone, water and air, that occur naturally, are called **raw materials** or **naturally occurring materials**. They are found in the Earth, in the sea or in the air. On the other hand, materials like iron and steel do not occur naturally, but they can be made from iron ore. Because of this, iron and steel are called **man-made materials**. Every year the chemical industry produces millions of tonnes of important man-made materials from raw materials. Table 1 shows the six most important raw materials and the substances we get from them.

	Raw material	Substances obtained from the raw material
1	Living things	Timber, foods (e.g. flour, oats, sugar, cooking oil, milk, meat, eggs, fruit, vegetables), clothing (e.g. cotton, linen), rubber
2	Coal	Fuels (e.g. coke, coal gas), dyes and plastics from coal tar
3	Oil and natural gas	Fuels, chemicals (e.g. plastics, pesticides, perfumes)
4	The sea	Salt (sodium chloride), sodium, chlorine, magnesium, bromine, water
5	The air	Oxygen, nitrogen, argon
6	Rocks in the Earth	Metals (e.g. iron, copper, aluminium, gold), limestone, sand, glass, cement, aggregate for building and construction

Table 1: Important raw materials and their products

Over 6 billion glass bottles and jars are used each year in Britain. Throwing these away produces 2 million tonnes of wasted glass

Conserving and recycling

Bottle banks are a common sight in Britain. In future we shall have to recycle other materials like plastics, aluminium and steel if we are to make the most of the Earth's resources

In our desire for more possessions and easier living we have taken more and more of the raw materials from the Earth. In doing this we have spoilt large areas of the Earth with mines, quarries, motorways and pylons. Large buildings have turned our cities into 'concrete jungles' and forests have been destroyed forever. As a result of this, important raw materials such as copper ore and oil, are being used up rapidly. Today there are 5000 million people on the Earth; there will be 7000 million by the year 2000. Unfortunately, *more* people need *more* food, *more* water, *more* raw materials and *more* land.

But *the Earth's resources will not last forever.* This fact has had some important effects.

- We are all more careful about the use of scarce resources. **Conservation** is essential.
- Waste materials, such as paper, plastics, glass, aluminium and steel, are being recovered and used again. This is called **recycling**.
- Scientists and engineers are producing cheaper **alternative materials** to those already in use.

Questions

1 How would you describe to a friend what chemistry is?

2 Make a list of 6 materials in your kitchen. If possible, give the chemical name of their main constituent (e.g. salt—sodium chloride). Say which materials occur naturally and which are man-made.

3 Coal is an important raw material. List 4 substances made or extracted from it.

4 (a) What does the word 'ore' mean?
(b) Which metals are obtained from the following ores: *bauxite, copper pyrites, haematite, galena, tinstone?*

5 Look at today's newspaper. What news or adverts does it contain about raw materials for the chemical industry?

Look at the variety of different materials in this room.

Most of the outside of this building is glass. What are the advantages of using glass? What are the disadvantages?

Look around your home. Notice the variety of different materials and their uses.

- **Metals** are used to make knives and forks, pans and ornaments. Metals have been used for these items because they are hard, strong, shiny and can be made into different shapes.

- Curtains, carpets and clothes are made from **fibres** which are soft and comfortable. Some fibres like cotton and wool occur naturally as part of plants and animals, but others like polyester (Terylene) and Nylon are made from the chemicals in coal, crude oil and plants.

- **Plastics** such as polythene, perspex and PVC are used for articles like clingfilm, washing-up bowls, combs and records. Plastics are used because they are fairly cheap, flexible (bendy) and easily moulded into different shapes.

- Windows, jars and some ornaments are made of **glass**.

- Window frames, doors, chairs and tables are often made of **wood**.

1 Why is glass used for windows?

2 Why is wood used for doors and furniture?

If you have tried to answer the last two questions, you will see that we choose different materials for different uses because they have different **properties**.

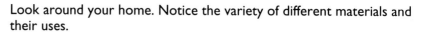

The properties of a material tell you what it is like and what it can be used for.

Classifying materials

Properties show clearly that one material is different from another. They also help us to put similar materials into groups or sets. For example, we can put drinks into different groups such as fizzy, milky, alcoholic and non-alcoholic. There is a special word for putting things into groups with similar properties. It is called **classification**.

Scientists classify materials in several ways. One of the most useful is as **solids, liquids and gases**. These are sometimes called the **three states of matter**.

The photographs and Table 1 below show some important properties of solids, liquids and gases.

Table 1: Important properties of solids, liquids and gases

Solids	**Liquids**	**Gases**
Solids have weight	Liquids have weight	Gases have weight
Solids have high densities usually greater than 2 gm cm^{-3}	Liquids have medium densities usually about 1 gm cm^{-3}	Gases have very low densities
Solids keep their shape	Liquids take the shape of their container (i.e. they can flow easily)	Gases take the shape of their container (i.e. they can flow easily)
Solids keep the same volume	Liquids keep the same volume	Gases change volume easily. They fill *all* of their container
Solids cannot be squashed (compressed) into a smaller volume	Liquids can be compressed very slightly	Gases can be compressed into a much smaller volume

The rocks at Stonehenge show the key properties of solids

The water flowing over Niagara Falls shows the key properties of liquids

The air inside this inflatable dinghy shows the key properties of gases

Questions

1 What properties of bricks and concrete make them important building materials?

2 At one time, gutters and drainpipes were made of iron. Today they are often made of plastics like PVC (polyvinyl chloride).
(a) What properties of iron made it useful for gutters and drainpipes?
(b) What were the disadvantages of iron?
(c) Why has iron been replaced by plastics?

3 Usually it is easy to tell whether a material is solid, liquid or gas. But sometimes it is more difficult.
(a) Is mud a liquid or a solid or does it contain both? Explain your answer.
(b) What state of matter is (i) smoke, (ii) jelly, (iii) custard, (iv) honey? Explain each of your answers.

4 (a) Make a table like the one below showing 4 different liquids that are used in a car and the jobs they are used for. The table has been started for you.

(b) Choose two of the liquids in your table. Say why they are used for their particular jobs.

Liquid used	What is it used for?
light oil	brake fluid
. . .	. . .

Gold is an element which occurs in the Earth's crust. This photograph shows the gold in Tutankhamun's funeral mask

The black material on badly burnt toast is carbon. No matter how this carbon is treated, it cannot be broken down into anything simpler

When bread is well-toasted, the surface gets covered in a black solid which is carbon. Smoke, containing water vapour and carbon dioxide, also rises from the burnt toast.

Changes like this, that result in new materials, are called **chemical reactions** and the new substance is called the **product** of the reaction. The chemical reaction which produced carbon from bread is caused by heat from the cooker. Many foods are prepared by chemical reactions caused by heat.

We can summarize the reaction which takes place when bread is heated by writing a **word equation**:

$$\text{bread} \xrightarrow{heat} \text{carbon} + \text{carbon dioxide} + \text{water} + \text{toast}$$

No matter how the black carbon is treated, it cannot be broken down into a simpler substance.

> *In 1661, Robert Boyle suggested the name **element** for a substance that cannot be broken down into a simpler substance. Elements are the simplest possible materials.*

Robert Boyle—the first scientist to use the word 'element'

This is still a useful description of an element. Carbon is an element. But substances like water and carbon dioxide are not elements because they can be broken down into simpler substances. Substances like water and carbon dioxide which contain two or more elements are called **compounds**.

When electricity is passed through water containing a little sulphuric acid, it breaks up forming hydrogen and oxygen (figure 1).

$$\text{water} \xrightarrow{\text{electricity}} \text{hydrogen} + \text{oxygen}$$

The hydrogen and oxygen cannot be made any simpler: they are elements.

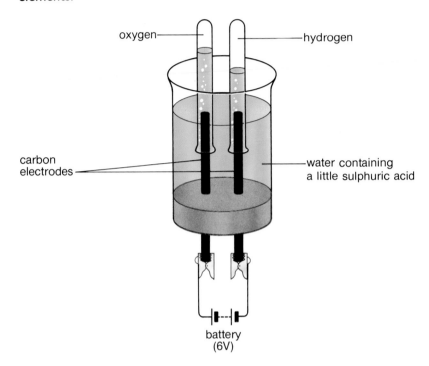

A lot of toys can be made from a few different pieces of Lego. In the same way, all materials are made from a few different elements

Figure 1
When electricity is passed through water containing a little sulphuric acid, it splits up into hydrogen and oxygen

There are millions of different materials in the universe, almost all of them are made through chemical reactions. All these materials are composed of elements. So far we know of 105 elements. These include aluminium, iron, copper, oxygen, nitrogen and carbon.

Although there are millions and millions of different substances in the universe, they can all be split up into one or more of the 105 known elements. For example, grass is made of carbon, hydrogen and oxygen. Sand is made of silicon and oxygen. So elements are the *building blocks for all substances*. In the same way, bricks are the building blocks for houses. Using a few different types of brick it is possible to build millions of different houses.

A reaction in which a compound splits up into two or more simpler substances is called **decomposition**. An example of this is the splitting of water into hydrogen and oxygen.

Water can be decomposed into hydrogen and oxygen, but if hydrogen is exploded with oxygen, these two simpler substances react to form water again. A reaction like this in which two or more substances join together to form a single product is called **synthesis**.

Notice that synthesis is the reverse of decomposition. Synthesis is a 'building up' whereas decomposition is a 'breaking down'.

$$\text{water} \underset{\text{synthesis}}{\overset{\text{decomposition}}{\rightleftarrows}} \text{hydrogen} + \text{oxygen}$$

Questions

1 Write a short article for 12 year olds explaining what elements are. Your article should include diagrams and photos if these will help.

2 Write down the names of all the elements mentioned in this unit.

3 A gas was produced when a solid was heated. How would you test to see if the gas is (i) carbon dioxide, (ii) water vapour, (iii) oxygen?

4 When red mercury oxide is heated, shiny beads of mercury are produced and oxygen gas is given off.
  (a) Write a word equation for the reaction.
  (b) Is this decomposition or synthesis? Explain your answer.

5 (a) Is it possible to decompose an element? Explain your answer.
  (b) Is it possible to decompose a compound? Explain your answer.
  (c) Which of the following can be decomposed?
  sugar, zinc, hydrogen, salt.
  (d) Explain why a synthesis reaction *always* produces a compound.

Why are metals ideal for making bells?

Element	Melting point /°C	Boiling point /°C	Electrical conductivity
Aluminium	660	2450	Good
Carbon (graphite)	3730	4830	Moderate
Copper	1083	2600	Good
Gold	1063	2970	Good
Iron	1540	3000	Good
Oxygen	−219	−183	Poor
Sulphur	119	445	Poor

Table 1: The properties of some elements

Look at the properties of the elements listed in table 1.

1 Which elements have a melting point above 500° C?
2 Which elements have a boiling point above 1000° C?
3 Which elements are good conductors of electricity?
4 Which of these elements can be polished to a shine?
5 Which of these elements can be bent or hammered into different shapes?
6 Which of these elements are metals?
7 Do the properties in Table 1 separate metals from non-metals?

Dividing elements into metals and non-metals is very useful. One of the best ways of checking an element is a metal or a non-metal is to see if it conducts electricity. This can be done by using the apparatus in Figure 1. The bulb lights when the element tested is a metal. When non-metals (except graphite) are tested, the bulb does not light because these are poor conductors.

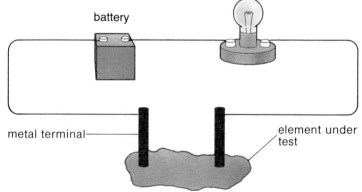

Figure 1

This photograph shows sulphur deposits. Sulphur is an important non-metal. It is used to manufacture sulphuric acid

Although one or two elements, such as graphite, are difficult to classify, the classification of elements as metals and non-metals is very helpful. There are about 80 metals, about 20 non-metals and five or six elements that are difficult to place. Table 2 compares the properties of metals and non-metals. There is more about classification in section E.

Marie Curie

Enrico Fermi

Albert Einstein

Dmitri Mendeléev

Which elements have been named after these four scientists?

Questions

1 Make a list of the properties of metals and non-metals.
2 (a) Make a table showing the names of 20 metals and 10 non-metals.
   (b) Write down the names of 2 elements that are difficult to classify as either metal or non-metal. Explain why they are difficult to classify.
3 How would you find out whether a chunk of solid element is a metal or a non-metal?
4 A colourless gas *A* relights a glowing splint:
   (i) *A* reacts with copper to give a black solid *B*;
   (ii) *A* reacts with a hot dark grey solid *C* to give a colourless gas *D*;
   (iii) *D* turns limewater milky;
   (iv) *B* and *D* are both produced when the green powder *E* decomposes on heating.
   (a) Identify the substances *A* to *E*.
   (b) Write word equations for the reactions (i), (ii) and (iv).

Property	Metals	Non-metals
State	Usually solids at room temperature	Solids, liquids or gases at room temperature
Melting point and boiling point	Usually high	Usually low
Density	Usually high	Usually low
Appearance	Shiny	Dull
Effect of hammering (malleability)	Can be hammered into shape (malleable)	Brittle (non-malleable) and soft when solid
Thermal and electrical conductivity	Good	Poor

Table 2: Comparing the properties of metals and non-metals

The bands of rocks in the Grand Canyon are mainly sandstones and limestones. The main elements in sandstones are silicon and oxygen. The main elements in limestone are calcium, carbon and oxygen

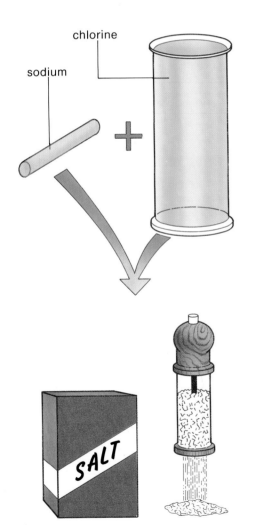

Figure 1
Sodium and chlorine are elements. They combine (react) to form salt (sodium chloride). Sodium chloride is a compound

When copper coins are first made, they look orange and shiny. Gradually they become dull as a layer of black copper oxide forms on the surface. The two elements, copper and oxygen have not just *mixed* together. They have joined together to form the new substance, copper oxide. We say that the copper and oxygen have **reacted** or **combined** to form copper oxide which is a compound. Many other elements can combine through chemical reactions to form compounds (figure 1).

> ● *A compound is a substance that contains two or more elements combined together.*

When *two* elements form a compound, the name of the compound ends in **-ide**. For example,

copper   +   oxygen   ⟶   copper ox**ide**
hydrogen +   chlorine   ⟶   hydrogen chlor**ide**

When a metal reacts with a non-metal, the non-metal forms the -ide part in the name of the compound. When two non-metals react, the more reactive non-metal forms the -ide part in the name of the compound.

Both elements and compounds can form mixtures. Most everyday materials are mixtures of compounds like sea water which contains salt (sodium chloride) and water (hydrogen oxide).

There are, however, some important mixtures of elements. For example, *metals can be mixed with other elements to form* **alloys**. Alloys have properties that are a combination of those of the elements they contain. So, it is possible to make alloys with specific properties. The most common alloy is steel. This is mainly iron with about 0.15% carbon. The carbon makes steel harder and tougher than pure iron. Stainless steel also contains chromium and nickel to prevent it from rusting. Iron, aluminium and copper are the most widely used metals. All three are used in large quantities in alloys.

Mixtures and Compounds

The important differences between mixtures and compounds can be illustrated using iron filings and sulphur. If you try this experiment, **wear eye protection**.

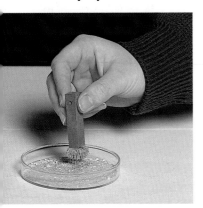

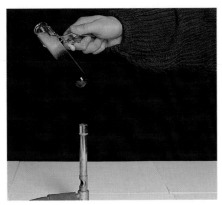

...king a mixture. Mix some iron filings with ...vdered sulphur. In the mixture, you can still ...the yellow sulphur. Iron can be separated ...m the mixture with a magnet

Making a compound. Heat a mixture of iron filings and sulphur. The mixture gets red hot and iron sulphide forms

When a mixture forms	When a compound forms
• Elements or compounds just mix together, they do not combine. • No composition of the mixture can vary. • No chemical reaction occurs. • The mixture has properties like the original substances. • The constituents can be separated by processes such as filtration, distillation and magnetic attraction.	• Elements or compounds combine (react). • A new substance forms. • The composition of the new compound is always the same. • A chemical reaction occurs. • The new compound has different properties from the elements in it. • The constituent elements in compounds cannot be separated easily.

Table 1: The differences between mixtures and pure compounds

When chemical reactions take place and new compounds are formed, there is no overall change in total mass. The mass of the new compounds (products) is just the same as that of the reactants. We say that mass is conserved and this important result is called the **Law of Conservation of Mass**.

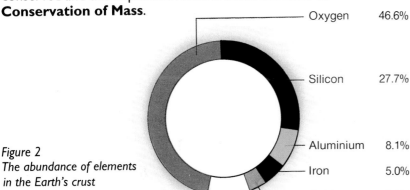

Figure 2
The abundance of elements in the Earth's crust

Oxygen 46.6%
Silicon 27.7%
Aluminium 8.1%
Iron 5.0%
Calcium 3.6%

Questions

1 Prepare a four-minute talk on alloys. In your talk:
  (a) explain what an alloy is,
  (b) give some examples of alloys,
  (c) show some uses of alloys,
  (d) say why alloys are important.

2 Mercury is the only liquid metal at room temperature. It has a high boiling point, a high density and it is a good conductor of heat and electricity. What uses does mercury have on account of these properties?

3 When water is added to white (anhydrous) copper sulphate in a test tube, a blue substance is produced. At the same time, the test tube gets hot.
  (a) Is the blue product a compound or a mixture?
  (b) Give 3 pieces of evidence to support your answer.

4 Write word equations for the reactions which occur when:
  (a) coke (carbon) burns in air (oxygen),
  (b) copper coins form a thin, dark layer of copper oxide,
  (c) hydrogen explodes with air (oxygen).

5 Figure 2 shows the abundance of the commonest five elements in the Earth's crust.
  (a) What total percentage do these five elements make up?
  (b) What percentage of the Earth's crust do the other 100 elements make up?
  (c) What is the most abundant metal in the Earth's crust?
  (d) Oxygen is the most abundant element. What materials in the Earth's crust contain oxygen?
  (e) Name one substance in the Earth's crust that contains
  (i) silicon, (ii) iron, (iii) calcium, (iv) aluminium.

6 Separating Mixtures

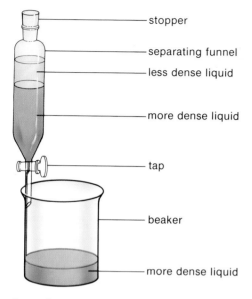

- stopper
- separating funnel
- less dense liquid
- more dense liquid
- tap
- beaker
- more dense liquid

Figure 1

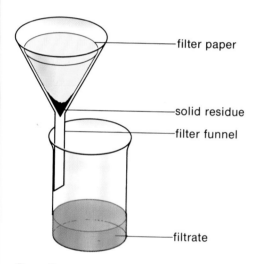

- filter paper
- solid residue
- filter funnel
- filtrate

Figure 2

Most substances which occur naturally are mixtures. Often these mixtures have to be separated before we can use them. Imagine what might happen if we used untreated, muddy water for cooking and cleaning, or if we tried to run our cars on crude oil rather than petrol.

Muddy water contains particles of solid 'floating' in it. If the water is left for some time, heavier particles of solid sink to the bottom as a **sediment**. The clearer water can then be poured off easily and separated from the sediment. This process is called **decantation**. Decantation is important in cooking. It is used to separate water from vegetables after cooking and also to separate two liquids which do not mix, like cream and milk. These liquids which do not mix are called **immiscible liquids**. Have you noticed how difficult it is to separate the cream from milk really well? The best way to separate two immiscible liquids is to use a separating funnel (figure 1).

If water is decanted from its sediment of mud, it still appears cloudy due to very small particles of solid 'floating' in it. We say that the particles are suspended in the water and that the mixture of fine solid particles and water is a **suspension**. In a suspension, some particles are so small that they do not sink. Smoke is a suspension of fine solid particles in air; milk is a suspension of tiny droplets of oil (cream) in a watery liquid.

In order to separate suspension of fine particles from cloudy water, we must use **filtration** (figure 2). There are millions of tiny holes in filter paper. Liquids pass through these holes, but solid particles are too large to do so. The solid that remains in the filter paper is called the **residue** and the liquid that trickles through is the **filtrate**.

Clean water for our homes

Filtration plays an important part in obtaining clean water for our homes. Figure 3 shows the main stages in the purification of river water.

The water is first stored in reservoirs where most of the solid particles can settle out. As the water is needed, it is filtered through clean sand and gravel which trap smaller particles of mud and suspended solids (figure 4).

After filtering, the water is treated with small amounts of chlorine to kill harmful bacteria in the water. The purified water is then pumped to storage tanks and water towers from which it flows to our homes.

Filtration is also used to separate beer from its sediment (yeast) before bottling. The beer is filtered by forcing it through filter cloths to catch the sediment.

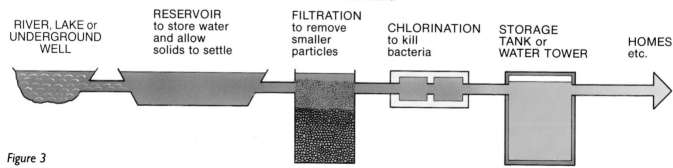

RIVER, LAKE or UNDERGROUND WELL — RESERVOIR to store water and allow solids to settle — FILTRATION to remove smaller particles — CHLORINATION to kill bacteria — STORAGE TANK or WATER TOWER — HOMES etc.

Figure 3

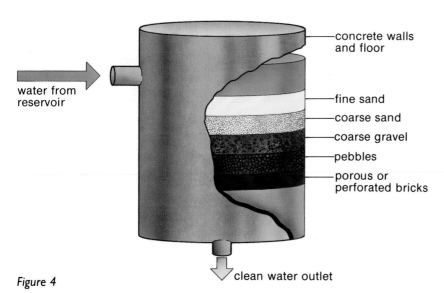

concrete walls
and floor

fine sand

coarse sand

coarse gravel

pebbles

porous or
perforated bricks

water from
reservoir

clean water outlet

Figure 4

Decanting cream from milk

Separating cream from milk

Another way to separate a solid from a liquid or to separate two immiscible liquids is to use a **centrifuge**. As the centrifuge spins round rapidly, a force pulls outwards and the heavier particles go to the bottom of the mixture. In the same way, a roundabout at the funfair drags you outwards. One of the most important uses of a centrifuge is in separating cream from milk to make skimmed milk. Whole milk is poured into a bank of spinning sloped discs. The heavier milk is forced down and outwards and the cream rises up and inwards (figure 5).

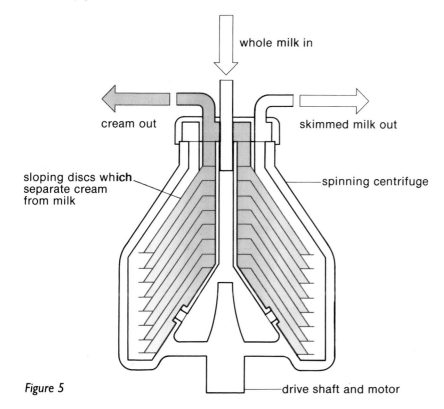

whole milk in

cream out

skimmed milk out

sloping discs which
separate cream
from milk

spinning centrifuge

drive shaft and motor

Figure 5

Questions

1 Explain the following:
immiscible liquids; filtrate; residue; suspension.

2 Give an example of separation by
(i) decantation;
(ii) filtration;
(iii) centrifugation.

3 Mist is a suspension.
(a) What are the suspended particles?
(b) What are they suspended in?

4 Draw a flow chart to summarize the main stages in the purification of river water for our homes.

5 Your younger brother and sister are playing at cooking. They decide to mix sugar with sand. Fortunately, you have a good chemistry set. Describe how you would separate dry sand and sugar crystals from their mixture.

6 You are given a yoghurt pot, some pebbles and some sand.
(a) Design a small-scale filtration plant which you could use to clean muddy river water. Draw a diagram of your design.
(b) How could you check how well your model works?

7 Separating Solutions

Sea-water is not pure—it contains salt and other dissolved substances essential for animals and plants that live in the water

Nail polish remover is a solvent for nail polish. The solvent is propanone (acetone)

Tap water is clean but *not* pure. It contains dissolved gases like oxygen and carbon dioxide and probably dissolved solids which make the water 'hard'. In hard water areas, the dissolved solids are left as scum after washing with soap. Clear sea-water is another example of a liquid containing dissolved solids. You can't see the salt in clear sea-water, but it must be there because you can taste it. The salt has been broken up into tiny particles which are too small to be seen even with a microscope. These particles are so small that they can pass through the holes in filter paper during filtration.

The mixture of dissolved salt and water forms a **solution**. The substance that dissolves is called the **solute**. The liquid in which the solute dissolves is the **solvent**. Solids such as salt and sugar which dissolve are described as **soluble** and solids such as mud which do not dissolve are **insoluble**.

Although water is a very good solvent, many substances are insoluble in it. Oils and greases are insoluble in water but very soluble in other solvents, like trichloroethane. This liquid is used by dry cleaners to remove grease stains from clothes. Nail polish is insoluble in water but it dissolves easily in a liquid called propanone (acetone). Nail polish remover is made of propanone.

Evaporation and crystallization

When sea-water is left to dry in the sun, the water turns into vapour (evaporates) and leaves behind salt. Next time you are at the seaside look for white rings of salt around the edges of rock-pools.

This change of liquid to a gas or vapour is called **evaporation**. Evaporation can be used to separate a dissolved solid from its solvent. If the solvent evaporates slowly, then the dissolved solute is often left behind as well-shaped crystals. We say that it has **crystallized**. In hot countries, salt is obtained from sea-water by allowing the water to evaporate, leaving the crystallized salt. Sugar is also obtained by a process of dissolving and evaporation. The sugar canes are crushed and then mixed with water to dissolve the sugar. The sugar solution is separated from the pulp and then water is evaporated off leaving sugar crystals.

At room temperature, salt solution takes a long time to evaporate and form crystals. We can speed up the process by heating the solution gently (figure 1). If the solution is evaporated quickly, the crystals are usually small. Larger crystals form if the solution is evaporated slowly. As the water evaporates, crystals begin to form at the edges of the solution. When this happens, it is best to stop heating and allow the solution to cool and crystallize slowly from then on. This gives larger, well-shaped crystals. If the crystals form in the solution, they can be dried between filter papers.

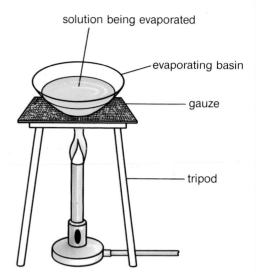

solution being evaporated

evaporating basin

gauze

tripod

Figure 1
If you try this experiment **wear eye protection**. *Tiny hot crystals may spit out of the evaporating basin*

Salt is harvested from sea-water in hot countries

Evaporation is an essential process in drying wet clothes and in producing concentrated evaporated milk. As the solvent evaporates from the solution, the same amount of solid is left in less and less solvent. We say that a solution which contains only a little solute in a given amount of solvent is **dilute**. If a solution contains a lot of solute in a given amount of solvent we say it is **concentrated**. As more and more solvent evaporates, a solution becomes more and more concentrated. Eventually, the solution cannot become any more concentrated. If any more solvent evaporates, some solid solute will come out of solution and form crystals. A solvent that has dissolved all the solute it can is **saturated**. As solvent evaporates from the saturated solution, more and more solid crystallizes out.

Questions

1 Explain what the following mean: *solution; solvent; solute; soluble; insoluble; saturated.*
2 Which of the following are solutions and which are suspensions? *ink; orange juice; steam; tea; exhaust fumes; beer.*
3 Which of the following are soluble in water? *sand; sugar; butter; aspirin; chalk; oil; vinegar.*
4 How would you obtain dry, well-shaped crystals from a solution of salt water?
5 How would you prepare a jug of filter coffee? In your answer use the words *solution, solvent, solute, dissolve, soluble, insoluble, filtrate.*

Whisky is made by distilling a liquid called wort in copper vessels. Wort is like weak beer. It is made by fermenting barley

Pure water from sea-water

We can get salt from sea-water by evaporating off the water. But can we get *pure water* from sea-water? When sea-water evaporates, water vapour escapes into the atmosphere. If the water vapour is passed into a second container and cooled, it will turn back to water. All the solute (salt) stays in the solution, so the water which evaporates and cools is pure water.

When a vapour changes to a liquid the process is called **condensation**. The apparatus which cools the vapour to liquid is a **condenser**. Figure 1 shows how pure water can be obtained from sea-water by evaporating the water and then condensing the water vapour. This process of evaporating a liquid and condensing the vapour is called **distillation**.

distillation = evaporation + condensation.

When the sea-water in figure 1 is heated, the water vapour which escapes passes through the inner tube of the condenser. This is cooled by the cold water flowing around it, so the water vapour condenses and drips from the lower end of the condenser into a receiver. The pure water which collects is called the **distillate**.

Distillation can be used to separate any solvent from its solution. In parts of the Middle East, where fuel is cheap, distillation is used on a large scale to get pure water from sea-water. Pure water (distilled water) can also be obtained by distilling tap water. Most chemical laboratories have their own distillation apparatus called a *still* which produces distilled water.

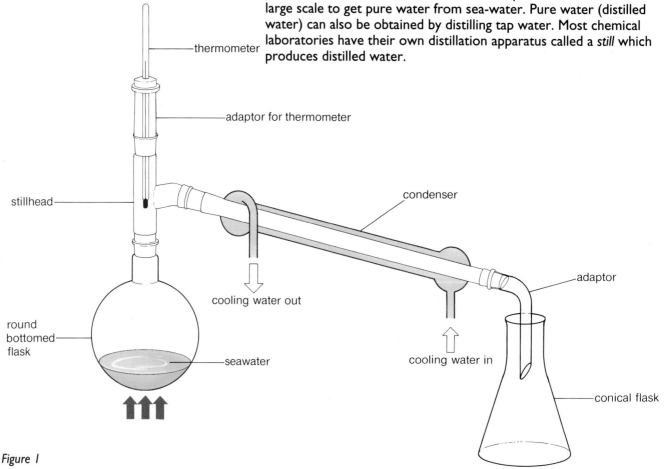

Figure 1

Petrol from crude oil

Crude oil is a mixture of many liquids and dissolved solids. Petrol and paraffin can be separated from it by distillation. When distillation is used to separate a mixture of liquids as in crude oil, it is called **fractional distillation** because it separates the mixture into two or more parts (fractions). Figure 2 shows the temperatures and products at different heights in an industrial fractionating column. Inside the column, there are horizontal trays with raised holes in them. The crude oil is heated by a furnace and the vapours then pass into the lower part of the fractionating column. As the vapours rise up the column through the holes in the trays, the temperature falls. Different vapours condense at different heights in the tower and are tapped off and used for different purposes. Liquids like petrol, which boil at low temperatures, condense high up in the tower. Liquids like lubricating oils, which boil at higher temperatures, condense lower down in the tower.

Fractional distillation is also used in industry to separate oxygen and nitrogen from the other gases in air (section B, unit 1).

Camping Gaz is obtained by fractional distillation of crude oil. It is the most volatile fraction in crude oil. Camping Gaz provides a portable fuel for cooking

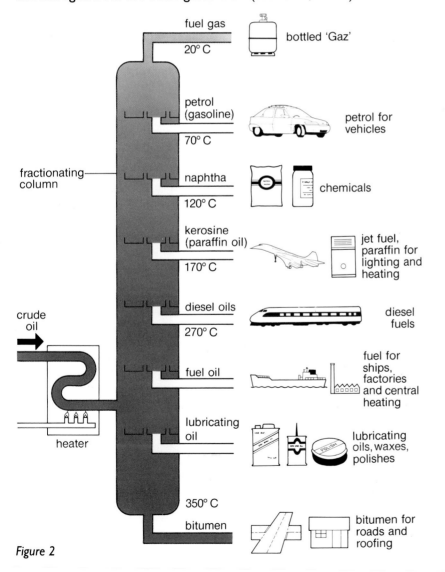

Figure 2

Questions

1 Explain the following:
evaporation; *condensation*; *distillation*; *distillate*.

2 What is the difference between simple distillation and fractional distillation?

3 Two important processes that involve distillation are the manufacture of brandy and whisky. Brandy is made by distilling wine; whisky is made by distilling beer. Both brandy and whisky contain alcohol (boiling point, 78° C) and water (b.pt. 100° C).

(a) Which liquid boils at the lower temperature, alcohol or water?
(b) Which of these two liquids evaporates more easily?
(c) If beer is distilled, will the distillate contain a larger or smaller percentage of alcohol?
(d) Why is whisky more alcoholic than beer?

9 Pure Substances

How pure is 'pure' honey?

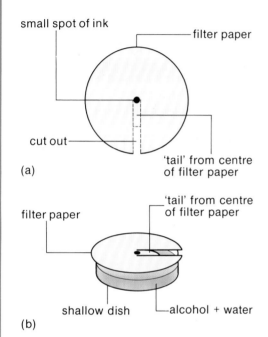

small spot of ink

filter paper

cut out

'tail' from centre of filter paper

(a)

'tail' from centre of filter paper

filter paper

shallow dish — alcohol + water

(b)

Figure 1

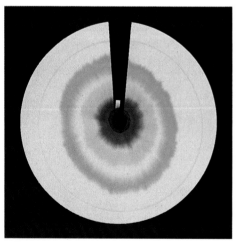

Figure 2

A chromatogram of black ink
(The solvent was butan-1-ol: ethanol: 0.88 ammonia solution 3:1:1 by volume)

The label in the photograph says that the honey is pure. This means that it has had nothing added to it or taken out of it. The so-called 'pure honey' contains many different substances. The proportions of these different substances gives each honey its own particular flavour.

> When chemists say that something is pure, they mean that it is a single substance and not a mixture of substances.

The honey is not pure to a chemist because it contains many substances. Most substances that occur naturally are mixtures, *not* pure substances. For example, air is a mixture of gases, mainly nitrogen and oxygen; petrol is a mixture of octane and other liquids and ink is a mixture of water and coloured dyes. If you have spilt ink on blotting paper you may have noticed that as the blot spreads out, it forms a series of different colours. Each of the coloured rings contains one of the dyes in the ink. Figure 1 shows how the dyes in ink can be separated more clearly.

Cut the filter paper as shown in figure 1(a) and fold back the tail from the centre. Put a spot of ink at the centre of the filter paper and leave it to dry. Then, place the filter paper on the shallow dish so that the 'tail' dips into the mixture of equal parts water and alcohol (figure 1(b)). As the liquid soaks up the 'tail' and through the filter paper, the different dyes in the ink spread out in a series of coloured rings (figure 2). The green dye in grass and the various coloured dyes in flower petals can be separated by a similar process. The coloured substances are first removed from the grass or petals by grinding in a pestle and mortar with alcohol.

This method was first used to separate coloured substances. The process was therefore called **chromatography** from the Greek word *khroma* meaning colour. Nowadays, the substances being separated may be colourless. After chromatography has taken place, the paper is sprayed with a liquid which reacts with the colourless substances to produce coloured substances.

Chromatography can be used to separate mixtures and to decide whether a substance is pure. It is very important in medicine because it can be used to separate drugs and other constituents in blood and urine.

The natural colours in fruit, vegetables and other foods can be improved by synthetic dyes. Chemists can detect these dyes using chromatography and check that they are harmless.

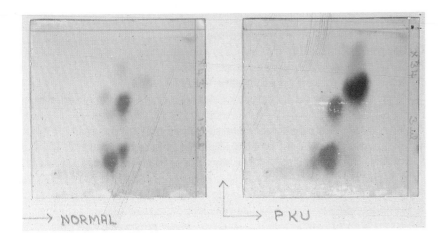

This photograph shows two chromatograms of substances called 'ketones' in the urine. The one on the left is from a normal child. The one on the right labelled 'PKU' is from a child suffering from phenylketonuria. This causes the child to excrete ketones in the urine, which show up as larger spots on the chromatogram

Testing for pure substances

There are three ways of testing to see if a substance is pure:
- using chromatography
- checking its melting point
- checking its boiling point

If a substance is pure it will contain only one kind of material. *So, if a coloured substance is pure, it will give only one coloured ring or band during the chromatography.*

The boiling point of a liquid changes as the atmospheric pressure rises or falls. On top of Mount Everest, where the pressure is much lower than at sea level, water boils at about 70° C. But, if the atmospheric pressure stays constant, *a pure substance will always boil at the same temperature*. In the same way *a pure substance always melts at the same temperature* if the pressure is constant. Because of this, we must pick a standard pressure and measure all boiling points and melting points at this standard pressure. Scientists have chosen the pressure exerted by a column of mercury 760 mm high as standard atmospheric pressure. This pressure is called 1 atmosphere (atm).

Pure water always boils at 100° C and pure ice always melts at 0° C at 1 atm pressure. Other substances melt and boil at different temperatures. So, measuring the melting point or the boiling point of a substance is the best way to decide if it is pure. If the substance is impure, its melting point and boiling point will differ from the values for the pure substance. For example, salty water boils above 100° C. As the water boils away and the solution gets more concentrated, the boiling point slowly rises.

Questions

1 What is a pure substance?
2 Describe how you would use chromatography to test whether some red ink contains a single pure dye or a mixture of dyes.
3 How would you check whether a colourless liquid is pure water? Draw a diagram of the apparatus you would use.
4 Which of the following are pure substances and which are mixtures? *sea water; petrol; iron; steel; steam; soil; milk; salt.*
5 Find out how chromatography is used either in medicine, or in testing athletes for illegal substances, or in checking the synthetic dyes in foods.

Section A: Activities

1 Making a cup of tea

Suppose you are going to make a cup of tea. Here are some rules that you should follow.

A Warm the teapot with boiling water and then pour the water away.

B Put one tea bag or one spoonful of tea in the pot for every two people.

C Pour freshly boiled water onto the tea.

D Let the tea infuse for five minutes and then stir.

E Serve the tea with milk and sugar to taste.

1 Why is it important to use boiling water?

2 What should the temperature of the water be when it is added to the teapot?

3 Why should the teapot be warmed before making the tea?

4 Why are tea leaves cut into small pieces before being used?

5 Some people prefer to use tea bags rather than loose tea leaves. Why is this?

6 What does the word 'infuse' mean in rule D?

7 Some people say that it makes no difference to the taste whether the tea is loose or whether it is in a bag. How could you test this? How will you make sure your test is fair?

8 Some people prefer to put milk in the cup before pouring in the tea. Other people put the tea in first and then add the milk. Plan an experiment to investigate whether pouring in the milk or the tea first makes any difference to the temperature.

9 If possible, try the experiments which you have suggested in parts **7** and **8**. What are your results?

2 Sparks fly in schoolgirl horror!

SPARKS FLY IN SCHOOLGIRL HORROR!

Schoolgirl, Mandy Clarkson, aged 15, described how her hair had gone up in flames during a science lesson. Mandy said, "Suddenly my hair was in flames. I am sure my hair did *not* touch the Bunsen flame. I think it must have been caused by the gel on my hair."

Fortunately, Mandy is not badly burnt, but she will have to wear a wig for the next few months.

Hair gels usually contain three ingredients

(i)   a solvent which evaporates easily,

(ii) glycerine (propane-1,2,3-triol) or a polymer to gel the mixture,

(iii) perfume.

The solvent is probably flammable, but it is volatile and should evaporate quickly. Accidents like the one reported in the newspaper cutting have occurred two days after the gel was used.

The only ingredient which is likely to remain on the hair after some time is the glycerine or the polymer. Neither of these is very flammable. They are however spread over the hair to give a large surface area. Even so, it is difficult to see why the hair should catch fire so easily. One possible explanation is that the gelling stops the solvent evaporating.

Suppose you are taking an assembly at school. Your teacher has asked you to talk for three or four minutes about the hazards of hair gels. Remember that students may be at risk at home and in other lessons besides chemistry. Make this clear in your talk and suggest precautions that they should take when using hair gel. Now, write your talk.

Iced tea

Lemon tea

3 | Lemon tea and iced tea

Lemon tea and iced tea are two interesting ways of drinking tea that you may not have tried.

Lemon tea is pleasant all the year round.

(i) Prepare the tea in the usual way by pouring freshly boiled water onto the tea leaves.
(ii) Allow the tea to infuse for five minutes and then stir.
(iii) Use a tea strainer to collect the tea leaves and then serve the tea with a slice of lemon.
(iv) Sugar or sweetener may be added to taste.

Iced tea is pleasant on a hot summer's day.

(i) Prepare the tea in the usual way.
(ii) After the tea has been infused for five minutes, use a tea strainer to collect the tea leaves and pour the tea off into a jug.
(iii) Put the jug of tea into the fridge to cool.
(iv) Serve with ice cubes, lemon or orange slices if you wish.

Copy the recipes for lemon tea and iced tea carefully.

1 Underline all the names of solutions in the recipes (e.g. lemon tea in the first line).
2 Put a circle around the names of any solvents.
3 Put a tick through the names of any solutes.
4 Put a cross through the names of any insoluble substances.
5 Draw a flow chart to show the stages in making a cup of lemon tea. Illustrate each stage with a diagram.
6 Why is lemon tea often recommended in slimming diets?

4 | Word processing your own word quiz

1 Make a list of the important words in section A, units 1 to 5. After each key word, write down what it means. For example:

Chemistry—The study of materials and substances.
Substances—Materials like iron, iron ore, limestone, coke and air.
Raw materials—Materials that occur naturally.

In the diagram opposite, these three important words have been used to make a word quiz.

Can you see that the missing words is AIR?

2 Now make up a word quiz of your own using important words from the whole of section A. Write clues to find your missing word as in the example above.
3 Use a word processor to design and develop a neat version of your word quiz. Try to use all the facilities of your word processor including different sized letters, drawings (diagrams), shading, brushwork, etc.
4 Try out your word quiz on your friends.

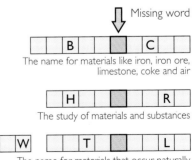

Missing word

| | B | | | C | |

The name for materials like iron, iron ore, limestone, coke and air

| | H | | | R | |

The study of materials and substances

| | W | | T | | L |

The name for materials that occur naturally

Section A: Study Questions

1 Some properties of six elements, *A* to *F*, are given below.

Element	Density /g cm⁻³	Boiling point /°C	Electrical conductivity
A	3.12	58	Poor
B	8.65	765	Good
C	19.30	2970	Good
D	3.4×10^{-3}	−152	Poor
E	0.53	1330	Good
F	2.07	445	Poor

(a) Which of the six elements are metals?
(b) Which element will float on water as a solid at 20° C?
(c) Which element is a gas at room temperature?
(d) Which non-metal may be a solid at 60° C?
(e) Which metal is probably a liquid at 2000° C?
(f) What is the boiling point of the least dense metal?
(g) The elements in the list are cadmium, gold, lithium, bromine, krypton and sulphur. Which element is which?

2 What elements do the following contain?
octane (in petrol), marble (calcium carbonate), anhydrous copper sulphate, sodium chloride, sugar, bread, cheese, wood, a penny, alcohol, polythene, water.

3 Seven processes used in chemistry to separate mixtures are:
A filtration, *B* distillation, *C* fractional distillation, *D* solution followed by filtration, *E* chromatography, *F* crystallization, *G* evaporation to dryness.

Choose from *A* to *G* the method that would be used
(i) to obtain copper(II) sulphate crystals from copper(II) sulphate solution,
(ii) to obtain pure water from sodium chloride solution,
(iii) to isolate nitrogen from liquid air,
(iv) to obtain calcium carbonate from a mixture of calcium carbonate and water,
(v) to obtain sand from a mixture of salt and sand,
(vi) to separate a mixture of alcohol and water,
(vii) to find out how many coloured substances are present in a purple dye.

4 The following chromatogram was obtained in an experiment to analyse two mixtures.
(a) (i) Why was a pencil used to mark the chromatogram and not a pen?
(ii) Why must the base line, on which a small drop of each sample was placed, be above the level of the solvent at the start?
(b) From the chromatogram, which of the single substances were
(i) present in mixture 1?
(ii) present in mixture 2?
(iii) not found in either mixture?
(c) The R$_f$ value of a substance can be used to identify it by chromatography. The R$_f$ value is defined as:

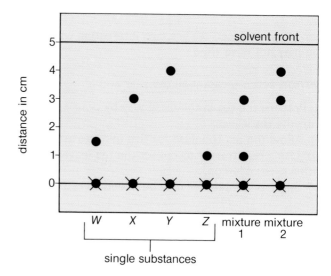

$$R_f = \frac{\text{distance moved by substance}}{\text{distance moved by solvent front in the same time}}$$

Calculate the R$_f$ value of substance Y.

5 A sample of crude oil was distilled in the apparatus shown in figure 1. Four fractions of distillate were collected in the temperature ranges:
20° C − 70° C, 70° C − 120° C, 120° C − 170° C and 170° C − 220° C
(a) The crude oil is not heated alone but with mineral wool. Why is mineral wool used?
(b) Why is the receiving tube in a beaker of cold water?
(c) Why would it be better to use a condenser to condense the vapours?
(d) Which fraction will have condensed least efficiently?
(e) Which fraction of the distillate will collect first?
(f) Which fraction of distillate could be used as petrol?
(g) Which fraction of distillate is most flammable?
(h) Which fraction burns with the smokiest flame?
(i) Which fraction would be darkest in colour?
(j) Why is the mineral wool black at the end of the experiment?

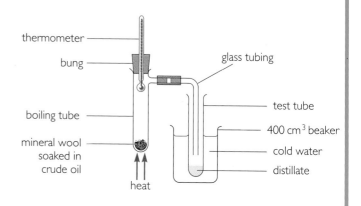

SECTION B

Our Environment—
Air and Water

Smoke pollution over Manchester and river pollution in the Grand Union Canal

23

1 The Air

The air is all around us. We need it to live. We need it to burn fuels and keep warm. We also obtain useful products from air. These products include oxygen and nitrogen. Air is a mixture, but the substance in it that we need for both burning and breathing is **oxygen**.

What percentage of the air is oxygen?

Figure 1 shows an apparatus to find the volume of air used up when copper reacts with the oxygen in it.

copper + oxygen → copper oxide

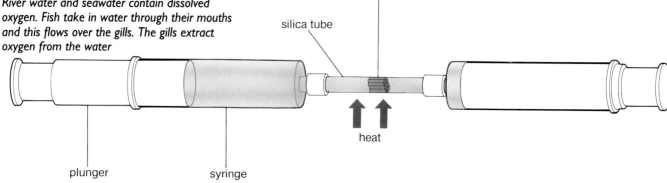

River water and seawater contain dissolved oxygen. Fish take in water through their mouths and this flows over the gills. The gills extract oxygen from the water

Figure 1
Wear eye protection *if you try this experiment*

At the beginning, one syringe is empty and the other is filled with 100 cm³ of air. The silica tube is heated strongly and the air is pushed to and fro several times so that all the oxygen in the air reacts with the copper. The tube is then allowed to cool and the volume of gas in the syringe is measured. The heating and cooling are repeated until the volume of gas which remains in the syringe is constant. The following table shows the results from one experiment.

Volume of air in syringe before heating	100 cm³
Volume of gas after first heating and cooling	82 cm³
Volume of gas after second heating and cooling	79 cm³
Volume of gas after third heating and cooling	79 cm³

Table 1: The results of an experiment to find the percentage of oxygen in air.

1 Has all the oxygen been used up after the first heating?
2 Has all the oxygen been used up after the second heating?
3 Why is the heating and cooling repeated three times?
4 How much oxygen did the copper remove?
5 What is the percentage of oxygen in the air?

Humans have no gills. They must carry a supply of air for underwater swimming

Obtaining gases in the air

Table 2 gives some information about the gases in dry air.

Gas	% of air by volume	Boiling point /° C	Important uses	
Nitrogen	78.09	−196	making ammonia for nitric acid and fertilizers	
Oxygen	20.95	−182	steel making and welding (see section B, unit 3)	
Argon	0.93	−186	filling electric light bulbs (Argon is very unreactive. It will not even react with the white hot filament.)	
Carbon dioxide	0.03	−78	in fizzy drinks and fire extinguishers	

Table 2

Notice that fresh air always contains about 0.03% carbon dioxide. This is because:

- carbon dioxide is produced when fuels burn and when plants and animals respire (see unit 3).
- carbon dioxide is used up when plants photosynthesize.

Although clean dry air contains only those gases in the table, ordinary air also contains water vapour and waste materials from industry (unit 4).

Obtaining gases from the air

The air is an important source of oxygen, nitrogen and argon. Figure 2 summarizes how the gases are separated.

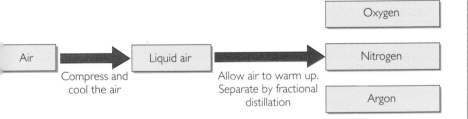

Air → Compress and cool the air → Liquid air → Allow air to warm up. Separate by fractional distillation → Oxygen / Nitrogen / Argon

Look at the boiling points of the gases in table 1.

1 Which gas forms a liquid first as the air is cooled?
2 Which gas boils off first when liquid air warms up?
3 Which gas boils off second when liquid air warms up?
4 Which gas boils off third when liquid air warms up?

Questions

1 In the syringe experiment to find the percentage of oxygen in the air, the hard-glass tube containing copper was weighed before and after the experiment.

Volume of oxygen in air	21 cm³
Mass of tube + contents at start	11.365 g
Mass of tube + contents finally	11.393 g

(a) Why does the tube + contents increase in mass?
(b) What volume of oxygen does the copper react with?
(c) What mass of oxygen does the copper react with?
(d) What is the density of this oxygen (i.e. the mass of 1 cm³ of it)?

2 Discuss the following questions with 2 or 3 others. The percentage by volume of carbon dioxide in ordinary air is 0.03%. Would you expect the percentage of carbon dioxide to be higher, lower or about the same as this
(a) in a crowded classroom?
(b) in a greenhouse full of plants on a sunny day?
(c) in the centre of a busy city?

2 Oxygen

Metal sculptors use oxy-acetylene welders to shape and cut their designs

The uses of oxygen

Oxygen is essential for burning and breathing, but the oxygen does not have to be pure. Pure oxygen is, however, required in large quantities for industrial and medical uses.

- **Manufacture of steel.** One tonne of pure oxygen is needed to produce every 10 tonnes of steel from impure iron ore. Oxygen is blown through the molten iron to remove carbon and sulphur impurities. The carbon and sulphur are converted to carbon dioxide and sulphur dioxide and then escape as gases.

- **Welding and cutting.** Pure oxygen is used in oxy-acetylene welding and cutting. When acetylene burns in oxygen, the temperature reaches 3200° C. This is hot enough to melt most metals which can then be cut or welded together.

- **Breathing apparatus.** Pure oxygen is used in life-support machines in hospitals. Oxygen helps the breathing of patients with lung diseases such as pneumonia. It is also mixed with anaesthetizing gases during surgical operations.
Mountaineers and deep sea divers also use oxygen mixed with other gases to breathe when supplies of air are not available.

Oxygen being used to revive someone

Element	Reaction with oxygen	Product	Add water to product, then universal indicator
Carbon	Glows red hot, reacts slowly	Colourless gas (carbon dioxide)	Dissolves pH = 5 acidic
Sulphur	Burns readily with a blue flame	Colourless gas (sulphur dioxide)	Dissolves pH = 3 acidic
Sodium	Bright yellow flame—white smoke and powder	White solid (sodium oxide)	Dissolves pH = 11 alkaline
Magnesium	Dazzling white flame—white clouds and powder	White solid (magnesium oxide)	Dissolves slightly pH = 8 alkaline
Iron	Glows red hot and burns with sparks	Black-brown solid (iron oxide)	Insoluble
Copper	Does not burn, but the surface turns black	Black solid (copper oxide)	Insoluble

Table 1: Comparing the reactions of some elements with oxygen

How does oxygen react with elements?

Things made of iron are slowly covered with rust (iron oxide) on exposure to the air. Similarly, shiny aluminium surfaces get covered with a layer of white aluminium oxide as the aluminium reacts with oxygen in the air. Table 1 shows the results obtained when various elements are heated strongly in oxygen.

All the elements burn better in oxygen than in air. The substances produced are **oxides**.

carbon + oxygen → carbon dioxide
sodium + oxygen → sodium oxide

> ● *The metal oxides (sodium oxide, magnesium oxide, iron oxide and copper oxide) are all solids.*
> ● *The non-metal oxides (carbon dioxide and sulphur dioxide) are both gases.*

How do oxides react with water?

Each of the oxides was shaken with water and the solution produced was then tested with universal indicator. The results are given in table 1.

Sodium and magnesium oxides react with water to form alkaline solutions with a pH greater than 7.

sodium oxide + water → sodium hydroxide
magnesium oxide + water → magnesium hydroxide

Iron oxide and copper oxide do not react with water or dissolve in it. So, they do not change the colour of the indicator.

> *The oxides of metals are called* **basic oxides**. *Most metal oxides are insoluble in water but a few, like sodium oxide and magnesium oxide, react with it to form alkaline solutions. These oxides are called* **alkaline oxides**. *Figure 1 shows the relationship between basic oxides and alkaline oxides in a Venn diagram.*
> *The non-metal oxides react with water to form acidic solutions with a pH less than 7. These oxides of non-metals which give acids in water are called* **acidic oxides**.

carbon dioxide + water → carbonic acid
sulphur dioxide + water → sulphurous acid

What is the order of reactivity?

Some metals, such as sodium and magnesium, react very vigorously with oxygen producing a bright flame. Others, like copper, react very slowly with no flame. These metals only form a thin layer of oxide. From these results we can arrange the metals in order of reactivity with the most reactive metal at the top and the least reactive at the bottom. This is called a **reactivity series** for the metals. Figure 2 shows a reactivity series for the four metals that we used.

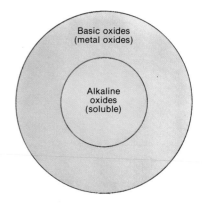

Basic oxides (metal oxides)

Alkaline oxides (soluble)

Figure 1

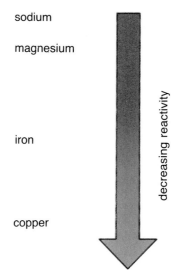

sodium

magnesium

decreasing reactivity

iron

copper

Figure 2

Questions

1 (a) What are the main uses of oxygen?
(b) Why do rockets carry liquid oxygen?
2 What is meant by the following: *basic oxide; acidic oxide; alkaline oxide?*
3 Barium burns in air to form a solid oxide.
This oxide reacts with water to give an alkaline solution.
(a) Is barium a metal or a non-metal?
(b) Write word equations for the reactions described above.
(c) Which of the following elements would react like barium? *copper; sulphur; sodium; iron; nitrogen; calcium; carbon; potassium.*

3 Burning and Breathing

The Greeks believed that Prometheus took pity on humans and stole fire from the gods to give warmth to men and women

Some fuels produce sparks and flames when they burn

Why is burning so important?

Burning was the first chemical process used by humans. Our ancestors burnt fuels to keep warm, to cook food and to produce new materials like metals and clay pots. Burning is just as important to us. We burn fuels to keep warm, cook food, drive motor cars and generate electricity. Any substance which burns in air to produce heat is a **fuel**.

> The most important fuels are coal, oil and natural gas. These fuels are mainly compounds containing carbon and hydrogen. During burning, these elements combine with oxygen in the air to produce carbon dioxide, water and heat.
>
> fuel + oxygen → carbon dioxide + water + heat

Reactions like this which give out heat are called **exothermic reactions**.

Sometimes, when a fuel burns, so much heat is produced that the products burst into flames. This happens in a Bunsen burner or a gas cooker when gases react with oxygen in the air to produce the flame. Reactions like these can be hazardous and must be carefully controlled.

Some fuels, like hydrogen, and explosives, like TNT and dynamite, react so fast that they cause explosions. Because of this *great care is needed in handling and using fuels and explosives.*

Breathing and respiration

When we breathe, air is taken into our lungs and then breathed out. But what happens to it in-between? One way to find out is to

Lavoisier showed that part of the air is used up when things burn. He said this part of the air was oxygen

compare the air breathed in (**inhaled**) with the air breathed out (**exhaled**). The temperature and the percentages of oxygen, carbon dioxide and water vapour in ordinary air and in exhaled air are shown in Table 1.

	Ordinary air (inhaled air)	Exhaled air
% oxygen	20	15
% carbon dioxide	0.03	5
% water vapour	2	5
temperature/° C	15 (variable)	25

Table 1: Look at these results. How does inhaled air compare with exhaled air?

Once the air is in our lungs, oxygen can get into our bloodstream. The oxygen then passes to the rest of our bodies where it reacts with substances in our food. This process is called **respiration**. The results in table 1 show that during respiration:

- oxygen is used up
- carbon dioxide is produced
- water is produced
- heat is produced

So respiration can be summarized as

food + oxygen→carbon dioxide + water + heat

Notice the similarity between this word equation and that for fuels on the previous page. Foods are special kinds of fuels which are used in our bodies. They contain carbon and hydrogen in different compounds like fats, carbohydrates and proteins. Respiration is a kind of very slow burning. The same amount of heat is produced whether a substance burns in the air with flames or whether it is used up slowly in our bodies as a food.

Some athletes chew glucose sweets before a marathon. Glucose is a carbohydrate. It is a very good supply of fuel during a long race.

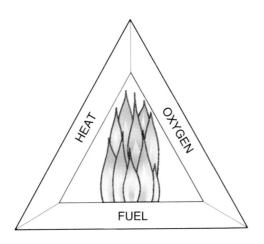

Questions

1 How do you think the fire service would deal with
(i) a petrol fire, (ii) a forest fire, (iii) a chip-pan fire?
Say what they would do and why.

2 (a) What fuels are used for each of the following?
barbecueing, cooking in a tent, heating a school, getting a spaceship into orbit, lighting a housefire, running a car.
(b) For each of the uses in part (a) say why the particular fuel is chosen.

3 Look at the photograph of the French scientist Lavoisier on page 28. In May 1794, during the French Revolution, Lavoisier was arrested on a trumped-up charge and guillotined. The judge pushed aside his defence with the words *'The Republic has no need of men of science.'* A friend of Lavoisier is reported to have said *'Only a moment to cut off his head and a hundred years may not give us another like it.'*
(a) What is meant by (i) trumped-up charge, (ii) The Republic?
(b) Find out about Lavoisier's experiments.
(c) What important results did Lavoisier obtain?
(d) How did Lavoisier's results help scientists to understand burning?
(e) Pretend that you were Lavoisier's lawyer at his trial. How would you reply to the judge when he said 'The Republic has no need of men of science'?

4 The diagram on the left shows the fire triangle.
(a) Why are heat, oxygen and fuel shown along the sides?
(b) What happens if one of these is missing?
(c) What must you do to stop a fire?
(d) Have you ever needed to stop a fire? If so, what did you do and why?

4 Air Pollution

Air pollution is worst in heavily industrialized areas. Smoke like this causes windows, curtains and clothes to become dirty very quickly. Acids in the fumes lead to the corrosion of metal structures and cause damage to the stonework of buildings

The two photographs show the Public Record Building, London before and after cleaning

Most air pollution is caused by burning fuels. The problems are therefore worst in industrial areas. When a fuel burns, it reacts with oxygen to form oxides. If the fuel burns completely, then all the carbon in it is turned into carbon dioxide which is only slightly acidic. If there is not much air available, the carbon may form **soot and smoke** or it may be turned into **carbon monoxide**, which is a very poisonous (toxic) gas. Carbon monoxide reacts with a substance in the blood called *haemoglobin*. This stops it carrying oxygen to the brain and other parts of the body. It is dangerous to run a car engine in a garage with the doors closed because the lack of air may lead to the production of carbon monoxide. Some fuels, like coal and coke, contain small amounts of sulphur. When these fuels burn, **sulphur dioxide** is produced. This is a colourless, toxic, choking gas which irritates our eyes and lungs. The sulphur dioxide dissolves in water to form an acidic solution containing sulphurous acid.

sulphur dioxide + water → sulphurous acid

When sulphur dioxide gets into rain, the rain water becomes acidic. This **acid rain** harms plants and attacks the stonework of buildings. Acid rain has been blamed for the poor growth of trees in Scandinavia and the death of fish in rivers and lakes. It is thought that the sulphur dioxide is produced in the industrial areas of Northern England and Southern Scotland and then carried by the prevailing winds across the North Sea to Scandinavia.

In a car engine, petrol burns in the cylinders. A lead compound is added to the petrol to help it burn smoothly, but waste lead compounds pass out with the exhaust gases. These **lead compounds** are poisonous because they affect the brain. In some countries including the UK, there are strict laws to control the amounts of lead compounds which can be added to petrol. All new cars are manufactured so that they can run on **unleaded petrol**.

Car engines need air to burn the petrol. When the mixture is sparked, nitrogen and oxygen in the air combine to produce **nitrogen dioxide** which is an acidic gas. Nitrogen dioxide causes acid rain like sulphur dioxide.

In the last few years, some scientists and politicians have become very concerned about the use of **CFCs (chlorofluorocarbons)** as refrigerator liquids and as the solvent in aerosol sprays. CFCs react with, and remove, ozone in the upper atmosphere. This allows more ultraviolet radiation in sunlight (normally absorbed by ozone) to get to the Earth. The extra ultraviolet radiation causes

1 damage to plants
2 greater risks of skin cancers.

CFCs, carbon dioxide and the other air pollutants also retain more heat than the gases in fresh air. This means that the temperature of the Earth is rising very slowly. This is called the **greenhouse effect**.

Acid rain has caused damage to this statue on Wells Cathedral, Somerset

The Tokyo Police are checking the exhaust gases of this car for pollutants. Tests on exhaust gases are now part of the MOT test in Britain

Advances in modern technology show both advantages and disadvantages. We have the benefits of more fuels and more vehicles, but we also have the problems of more pollution. Fortunately, several countries now have strict laws to control air pollution. In 1956, the Clean Air Act in Britain made it an offence to pollute the air with soot and smoke from factories and homes. Local authorities were given the power to set up smoke control areas ('smokeless zones') to prevent air pollution. Only smokeless fuels, such as coke and anthracite, can be used in smokeless zones.

Questions

1 (a) What are the main substances that cause air pollution?
  (b) How do they get into the air?
  (c) Suggest 3 ways in which air pollution causes damage.
2 (a) What is acid rain and how is it caused?
  (b) How does acid rain affect (i) lakes; (ii) forests?
3 What is (i) the Clean Air Act (ii) a 'smokeless zone'?
4 (a) What further steps could be taken to reduce air pollution in heavily industrialized areas?
  (b) What problems are there with stricter controls over air pollution?
5 Design a poster to show the dangers of CFCs.
6 It is the year 2030. You have just been elected as the first British Prime Minister to belong to the Green Party. Write down what you would say in your first interview on television about your policy on the environment.

5 Water Supplies

Water is the commonest liquid. We drink it, wash in it, swim in it and complain when it rains. In many ways, water controls our lives. It determines where we can live. It determines whether we can grow crops and produce enough food. It determines which sports we can enjoy. It determines the weather we have. All living things need water—people, animals and plants. About two thirds of your body is water. Everyday you need about 2 litres (3½ pints) of water. This water may be part of your food or drinks. It replaces the water that you lose in urine, in sweat and when you breathe. Water is more important than food—most people can survive for 50 to 60 days without food, but only 5 to 10 days without water.

Using water

Most of the time, we take water for granted. It is only during a very dry summer that we need to use it carefully. In the UK, each person uses about 180 litres of water every day. In some parts of Africa, each person must survive on less than 10 litres per day. Every time you flush the toilet about 10 litres of water are used and taking a bath may use 100 litres. Figure 1 shows the amounts of water that we might use for different purposes each day.

Large amounts of water are also used in industry. Most of the water is used for cooling. A large power station uses about 5 million litres of water per day. The other large industrial uses of water are for washing and as a solvent. When all these uses are added together, it takes:

10 litres of water to make 1 litre of lemonade or beer;
200 litres of water to make 1 newspaper;
50 000 litres of water to make 1 average-sized car.

Where does water come from?

At the turn of a tap we expect to get as much clean water as we need. But where does this treated water originally come from? If we trace the flow of water backwards, we find that it comes from:

- **surface water** in rivers and lakes or
- **ground water** in underground wells and saturated rocks.

We can trace the water supply back even further and see that all surface water and underground water comes from rain. When rain falls on the ground, some of its soaks deep into the earth as ground water. The rest runs off the land into rivers and lakes as surface water. This surface water covers 70% of the Earth. But where does the rain come from?

Water controls our lives. It affects our crops, our food, our health and the weather. 60% of the people in Asia and Africa cannot obtain clean water easily

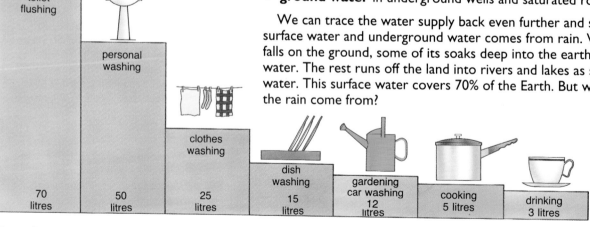

toilet flushing	personal washing	clothes washing	dish washing	gardening car washing	cooking	drinking
70 litres	50 litres	25 litres	15 litres	12 litres	5 litres	3 litres

Figure 1
Amounts of water used by one person each day in the UK

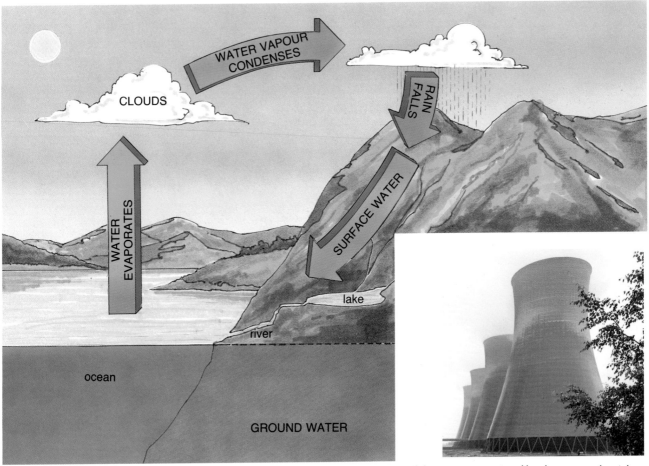

Figure 2 The water cycle

A large power station, like the one on the right uses 5 million litres of water every day for cooling. Because of this, power stations are often built near a large source of water, such as the sea, or a major river

Heat from the sun causes surface water to evaporate into the air from rivers, lakes and oceans. This water vapour then collects in the air as clouds. As the clouds rise, they cool down and the water vapour condenses to form drops of water. When these drops become large enough, they fall back to the earth as rain. The rain either soaks into the soil or joins rivers and oceans. The whole cycle then begins again. This continuous movement of water from the earth's surface to clouds and then back to the earth as rain is called the **water cycle** (figure 2).

Questions

1 Look at figure 1.
(a) Which of the uses of water requires (i) the purest water; (ii) the least pure water?
(b) Write down the uses, in order, from those needing the purest water to those needing the least pure water.
(c) Which of the uses could be avoided in order to save water during a drought?

(d) Suppose you saved all the water from personal washing. To which of the other uses could this water then be put?
2 (a) Make a list of all the things that your family uses water for.
(b) Estimate the amount of water, in litres, which your family uses for each of these, in one week.
(c) Find out how much your family pays in water rates.

(d) Calculate how much your family pays for 1 litre of water.
(e) Do you think that you get value for money?
3 (a) Draw a flow diagram to show how your tap water originally comes from rain.
(b) Where is the water treatment plant nearest to your home?
(c) What other services does your water company provide besides supplying water?

6 Water Pollution

This aerial photograph shows sea-water being badly polluted as a result of material being dumped at sea

Main sources of water pollution
Sewage
Fertilizers
Industrial chemicals
Pesticides
Oil
Detergents

Table 1: The main sources of water pollution

There are vast amounts of water in oceans, lakes and rivers, but it is easily polluted. The main sources of water pollution are listed in table 1.

Sewage is the main cause of water pollution. At one time, all sewage was pumped into rivers and the sea. This caused health hazards and led to diseases such as cholera. Sewage also affects living things in the water. If the amount of sewage is small, then bacteria in the water can break it down to harmless materials like carbon dioxide, nitrates and water. But, if the amount of sewage is large, then the bacteria use up all the oxygen dissolved in the water as they feed on the sewage. Once the oxygen concentration gets too low, most of the living organisms in the water (including the bacteria) die. The decaying materials make the water cloudy and smelly. In order to avoid this pollution, sewage plants treat the waste before it is returned to rivers and the sea. The sewage is pumped into large tanks and mixed with air (aerated) so that it can be decomposed more rapidly by bacteria.

During the 1950s, the River Thames was almost lifeless between London Bridge and the sea. This was caused by the low oxygen concentration in the water. Since then, there have been improvements in pollution control. In 1974, the first salmon was caught in the Thames since 1835 and now more than 70 species of fish have returned. In the last few years, Water Authorities have had to work to stricter levels of pollution control and this has led to cleaner rivers and beaches in Britain. Table 2 summarises the sources and effects of other substances which cause water pollution.

Some people believe that heat is also a serious water pollutant. It is a particular problem near the cooling water outlets from large nuclear power stations where the water temperatures may be 10° C higher than normal. The higher water temperatures increase the corrosion of steel structures in the water and also reduce the amount of oxygen dissolved in the water. The reduced oxygen content can be harmful to some fish such as salmon and trout, although other species thrive in the warmer water.

Polluting substance	Source	Effect
Fertilizers	Rain washes fertilizers into rivers and lakes	Bacteria and algae grow faster, use up all the dissolved oxygen and die
Industrial chemicals	Oils, metal compounds, acids, alkalis, dyes, etc. from factories	Toxic to animals, plants and bacteria in the water
Pesticides	Spraying of crops with chemicals	Toxic chemicals accumulate in the bodies of larger animals
Oil	Oil from refineries and ship-wrecked tankers	Covers sea birds with oil; pollutes beaches
Detergents	Factories, offices, homes	Causes water to foam; toxic to organisms in the water

Table 2: The sources and effects of various water pollutants

This guillemot and these razorbills have been killed by oil pollution at sea

This beach in the Gulf of Aqaba has been badly polluted from a factory which processes phosphate rock (calcium phosphate)

Getting the balance right

It is easy to adopt a simple view of pollution and say, 'Pollution is wrong'. But, if we did this and passed laws to protect our rivers and lakes, then industries would spend more money on avoiding pollution and removing waste in some other way. These additional costs would be passed on to the customers and goods from our factories would be more expensive. As a society we must decide between more pollution control and higher prices or less pollution control and lower prices. Remember, that *everyone* is responsible for pollution. We are all members of a society which consumes goods and materials and requires oil, fertilizers, pesticides and detergents.

Questions

1 What are the main sources of water pollution?

2 (a) Why does sewage cause fresh water to become murky and smelly?
(b) Why are bacteria mixed with sewage at some treatment plants?
(c) Why is solid sewage sometimes heated to 30° C and aerated regularly at the sewage plant?

3 (a) Suppose that each person in your family uses 150 litres of water every day.
(i) How much water is used by your family in one day?
(ii) How much water is used by your family in one year?
(b) Suppose your family pays £100 in water rates for the supply and treatment of water each year.
(i) What is the cost of water per litre for your family?
(ii) Compare this with the cost of one litre of petrol or milk.

7 Properties of Water

Living things, like this seal, can survive in water even in the Antarctic because ice floats on water

Water covers about 70% of the Earth's surface in oceans, lakes and rivers. It freezes at 0° C and boils at 100° C so it is a liquid at most places on the Earth. This is unusual. Can you think of another substance which occurs naturally as a liquid?

Water is also a very good solvent. It will dissolve substances as different as salt, sugar, alcohol and oxygen. Some substances are more soluble than others as you can see in the table.

Solute	Mass of solute which dissolves in 100 g of water at room temp. (20°C)
Sand	insoluble
Salt	36.0 g
Sugar	204 g
Alcohol	infinite
Oxygen	0.004 g
Carbon dioxide	0.014 g

Table 1: The solubility of some substances in water at 20°C

When the temperature falls below 0° C, water turns to ice. If the water is trapped in pipes as it freezes, then the pipes may split open and cause a burst. The pipes split because water expands when it freezes and the ice takes up more space than the water. This is another unusual property of water. Almost all other substances contract when they freeze. The expansion of water as it changes to ice means that ice is less dense than water. This expansion causes rocks to crack and potholes to appear in the roads.

When ice forms on the surface of a pond, it acts as an insulator and it prevents the water below from freezing. Because of this, living things can survive in water even in Arctic conditions.

Water as a solvent-solubility

When you stir a spoonful of instant coffee into a cup full of *hot* water, it dissolves very quickly. If you tried to make a cup of coffee with *cold* water, the coffee would not dissolve. You may have noticed that sugar dissolves better in hot coffee than in cold coffee. These everyday examples show that solids dissolve better in water as the temperature increases.

Water expands as it freezes. This can cause serious damage such as burst pipes

When sugar is stirred into coffee, it dissolves until a certain amount has been added. If more sugar is then added, this does not dissolve, provided the volume of the liquid and its temperature do not change. The coffee is **saturated** with sugar. The extent to which a solute like sugar can saturate a solvent like coffee is expressed in terms of **solubility**.

The solubility of a solute in a particular solvent is the mass of solute that saturates 100 g of solvent at a given temperature. Table 1 shows the solubilities of various substances in water at 20° C. The solubility of most solids increases with temperature.

Although most solids become *more soluble* in water as the temperature rises, gases become *less soluble* as the temperature rises. Natural water in lakes and rivers contains dissolved gases from the air such as oxygen and carbon dioxide. More gas dissolves if the water is colder or if the water flows over waterfalls where it mixes more freely with air.

Many animals and plants depend on oxygen dissolved in the water. Fish die if there is less than 0.0004 g of dissolved oxygen per 100 g of water. This explains why fish caught in rivers cannot survive in fish tanks indoors where the water is warmer with a lower concentration of dissolved oxygen.

Notice in Table 1 that carbon dioxide is more soluble than oxygen in water. This is because it reacts with water to form carbonic acid:

$$carbon\ dioxide + water \rightarrow carbonic\ acid$$

The carbonic acid which forms in natural waters reacts with chalk and limestone to form hard water. We shall return to this in a later unit.

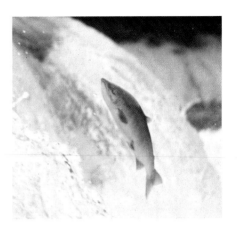

Some fish species require more dissolved oxygen in the water than others. Because of this they cannot survive at higher temperatures. Salmon require higher concentrations than most fish. They cannot survive if the water temperature is above 15° C

Questions

1 Why is water an unusual substance?

2 The solubilities of potassium nitrate and potassium chloride are given below at temperatures between 0°C and 60°C.

  (a) Draw the solubility curves of the two solids by plotting solubility on the vertical axis and temperature on the horizontal axis.

  (b) Which is the more soluble solid at 28°C?

  (c) At what temperature are they equally soluble?

  (d) How much potassium nitrate will crystallise from 100 g of water saturated with this substance at 50°C, if the temperature falls to 23°C?

3 Discuss the following questions in groups of 3 or 4.

  (a) Why is it necessary to give the temperature at which a solubility is measured?

  (b) How could you make sure that a solution is saturated at a particular temperature?

  (c) Why do you think that solids get more soluble in water as the temperature rises, but gases get less soluble?

  (d) Why does the water in fish tanks need aeration (air bubbled in)?

4 Design an experiment to show that water expands when it freezes.

Temperature/°C	0	10	20	30	40	50	60
Solubility of potassium nitrate /g per 100 g water	13	21	31	47	63	83	106
Solubility of potassium chloride /g per 100 g water	28	31	33	36	39	42	45

8 Rusting

Rusted old cars

Articles made of iron and steel rust much faster if they are left outside in wet weather. A garden fork rusts more quickly when it is left out in the rain than when it is kept dry in a shed. This suggests that water plays a part in rusting. But what other substances are involved?

When aluminium is exposed to the air, it becomes coated with a layer of oxide. The metal has reacted with oxygen in the air. If rusting is similar to this, iron may also react with oxygen during rusting.

The apparatus in figure 1 can be used to decide whether water and oxygen are involved in rusting. The test tubes are set up and left for several days. Tube 1 is the *control experiment*. It is the standard which we use to compare the results in the other tubes. It contains iron nails in moist air. Tube 2 contains iron nails and anhydrous calcium chloride to absorb water vapour and keep the air dry. Tube 3 contains nails covered with boiled distilled water. The water has been boiled to remove any dissolved air. The layer of olive oil prevents air dissolving in the water.

The nails in tube 1 and tube 4 rust, but those in tubes 2 and 3 do not.

1 Does iron rust if there is
   (i) no water;   (ii) no air;   (iii) no oxygen?
2 What conditions are necessary for rusting?

What is rust?

> *Iron will only rust if both*
> ● *oxygen and*
> ● *water are present.*

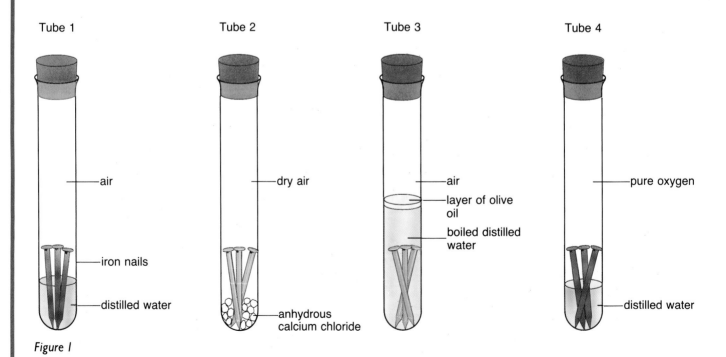

Figure 1

During rusting, iron reacts with oxygen to form brown iron oxide:

$$\text{iron} + \text{oxygen} \rightarrow \text{brown iron oxide}$$

At the same time, the iron oxide combines with water to form hydrated brown iron oxide. This is rust:

$$\text{iron oxide} + \text{water} \rightarrow \text{hydrated iron oxide (rust)}$$

Substances, like rust, that have water as part of their structure, are described as **hydrated** and we call them **hydrates**.

Blue copper sulphate crystals are also hydrated. They form when water is added to white anhydrous copper sulphate:

$$\begin{array}{lll}\text{anhydrous copper sulphate} + \text{water} \rightarrow & \text{hydrated copper sulphate} \\ \text{(white)} & \text{(blue)}\end{array}$$

The water present in hydrates is called **water of crystallization**.

Rusting costs millions of pounds every year because of

● the need to protect iron and steel objects;

● the replacement of rusted articles.

Although iron and steel rust more easily than several other metals, steel is used for ships, cars, bridges and other structures because it is cheaper and stronger than other building materials.

These car valves are being dipped in a solvent to prevent them from rusting

Large structures such as the Forth Rail Bridge are being painted all the time. As soon as the painters have finished they begin again at the other end. It would take one man 72 years to paint the Forth Rail Bridge

How is rusting prevented?

In order to stop iron and steel rusting, we must protect them from water and oxygen. The most important ways of doing this are:

● **Painting.** This is the usual method of preventing rusting in ships, vehicles and bridges.

● **Oiling.** The moving parts of machines cannot be protected by paint which would get scratched off. Instead they are oiled or greased (and this also helps lubrication).

● **Alloying.** Iron and steel can be mixed with other metals to form alloys. Stainless steel contains chromium, nickel and manganese mixed with iron.

● **Covering with a non-rusting metal.** Buckets and dustbins are coated ('galvanized') with a layer of zinc. Other articles, like car bumpers, taps and kettles, are chromium plated.

Questions

1 (a) What is rust?
  (b) How does rust form?
2 What methods are used to stop rusting?
3 Explain the following:
control experiment; hydrates; anhydrous; water of crystallization.
4 Explain the following statements.
  (a) Iron objects rust away completely in time.
  (b) Iron railings rust more quickly at the bottom.
  (c) Iron on shipwrecks in deep sea-water rusts very slowly.
5 What experiments would you do to find out whether:
  (i) iron rusts more quickly in sea-water or in distilled water;
  (ii) steel rusts more quickly than iron?
Draw a labelled diagram of the apparatus you would use in each case. What results would you expect?

9 Oxidation and Reduction

An unusual redox process. What substances are being oxidized and reduced?

When food goes bad, substances in the food combine with oxygen in the air. The process involves oxidation. Carbon dioxide and water are produced

Many reactions which occur in everyday life involve substances combining with oxygen to form oxides. Burning, breathing and rusting are three important examples.

During burning, fuels containing carbon and hydrogen react with oxygen to form carbon dioxide and water (hydrogen oxide):

$$\text{fuel} + \text{oxygen} \rightarrow \text{carbon dioxide} + \text{water}$$

During respiration, foods containing carbon and hydrogen react with oxygen to form carbon dioxide and water:

$$\text{food} + \text{oxygen} \rightarrow \text{carbon dioxide} + \text{water}$$

During rusting, iron reacts with oxygen and water to form hydrated iron oxide:

$$\text{iron} + \text{oxygen} + \text{water} \rightarrow \text{hydrated iron oxide}$$

Chemists use a special word for reactions in which substances combine with oxygen. They call the reaction **oxidation** and the substance is said to be **oxidized**. But if one substance combines with oxygen, another substance (possibly oxygen itself) must lose oxygen. We say that substances which lose oxygen in chemical reactions are **reduced** and we call the process **reduction**.

Oxidation and reduction always happen together. If one substance combines with oxygen and is oxidized, another substance must lose oxygen and be reduced. We call the combined process **redox** (*RED*uction + *OX*idation).

Notice that some redox reactions, like burning and respiration, are very useful, but other redox reactions like rusting and the spoiling of food, are a problem.

Redox in industrial processes

A redox reaction is sometimes used in welding steel. A mixture of powdered aluminium and iron oxide (known as 'thermit') is used. Aluminium is more reactive than iron, so it removes oxygen from iron oxide to form aluminium oxide and iron. (Figure 1).

Aluminium gains oxygen and is oxidized. Iron oxide loses oxygen and is reduced. The whole process involves both oxidation and reduction. It is an example of redox.

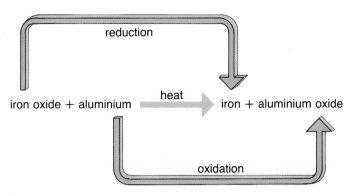

Figure 1

This reaction between aluminium and iron oxide is very exothermic. The iron is produced in a molten state. As the iron solidifies, it will weld two pieces of steel together. Welding together Inter-City railway lines using this method has increased the average 'life' of the track from 23 to 30 years. The photo (below) shows the process in action.

Redox reactions also occur in the manufacture of metals. Aluminium is manufactured by passing electricity through aluminium oxide. The electricity decomposes the aluminium oxide into aluminium plus oxygen. In this process aluminium oxide is reduced to aluminium.

$$\text{aluminium oxide} \xrightarrow{\text{electricity}} \text{aluminium} + \text{oxygen}$$

In a blast furnace, iron oxide (iron ore) can be reduced to iron with coke (carbon). The overall reaction in the manufacture of iron from iron ore is:

iron oxide   + carbon   →   iron + carbon dioxide
(iron ore)   (coke)

Which substance is oxidized and which substance is reduced in this process?

The redox reaction between aluminium and iron oxide can be used to weld railway lines together. Molten iron from the reaction runs into a mould around the rails to be joined. When the iron has cooled the mould is removed and excess metal trimmed off

Questions

1 Look at the food going bad in the photo on page 40.
(a) What chemical processes occur when food decays?
(b) What methods and precautions do we use at home to stop food going off?
(c) How do these methods work?

2 Which substance is oxidized and which is reduced in each of the redox reactions below?
(a) aluminium + water → aluminium oxide + hydrogen
(b) hydrogen + oxygen → water
(c) copper oxide + hydrogen → copper + water

3 (a) Will copper oxide and magnesium react on heating? Explain your answer.
(b) Will magnesium oxide and copper react on heating? Explain your answer.
(c) Element *W* reacts with the oxide of element *X* but not with the oxide of element *Y*. Write *W*, *X*, and *Y* in order of reactivity (most reactive first).

4 Find out how either chromium or titanium is manufactured. How is redox involved in these processes?

5 A sample of dry rust was heated strongly in a test tube. Droplets of a liquid, *A*, formed on the cooler parts of the tube. *A* turned anhydrous copper sulphate blue. The residue left in the tube, *B*, was heated strongly in a stream of gas, *C*. A grey-black solid, *D*, was formed (which conducted electricity) and a colourless liquid, *E*, condensed on the cooler parts of the apparatus. *E* turned anhydrous copper sulphate blue.
(a) Identify the substances, *A–E*.
(b) One of the reactions described above involves redox.
(i) Which is it?
(ii) Write a word equation for the reaction involved.
(iii) Explain which substance is oxidized and which is reduced.

Section B: Activities

1   Fuels in the home

1 Make a survey of the methods of central heating and the fuels used in the homes of the people in your class. You will need to find the following information:

(i) the number of homes in your survey,

(ii) *central heating*—how many homes use:

- gas-fired central heating
- oil-fired central heating
- solid-fuel (coke, coal) central heating
- electricity for central heating (i.e. electric storage heaters)

(iii) *other fuels*—how many homes use:

- electric fires
- fan heaters
- gas fires
- wood

- coal fires
- paraffin heaters
- smokeless fuel
- solar heaters

2 Draw a bar chart showing the number of homes using the different forms of central heating listed in part (ii).

3 Draw a bar chart showing the number of homes using the other fuels listed in part (iii).

4 How do you think the number of homes using the different forms of central heating has changed in the last 20 years?

5 How do you think the number of homes using the different forms of central heating will change in the next 20 years?

2   Fire fighting

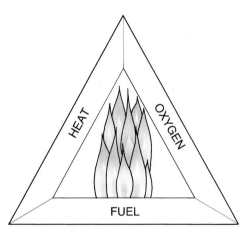

The fire triangle

Fire-fighters continue to damp down a smouldering car to ensure that it is no longer a fire hazard

The fire triangle shown on the left can be used to explain how fires start and how they can be put out.

1 Why does the fire triangle have labels for fuel, oxygen and heat?

2 When the fire service are called to a fire, they try to remove either the fuel, the oxygen, the heat or more than one of these three. Why is this?

3 Which one of fuel, oxygen or heat is being removed from the fires described below?

(a) When the fat in the chip pan suddenly caught fire, Nicola grabbed a damp cloth and covered the chip pan.

(b) When the forest fire started, the foresters cleared the fire breaks (i.e. the wide tracks in the forest) just ahead of the fire.

(c) The firemen sprayed water on the burning beams of the warehouse.

(d) Julian was sitting close to the fire. Suddenly his coat was in flames. Shakila acted quickly and rolled Julian in the rug on which he was sitting.

3 | Pollution of the River Rhine and the North Sea

The reports

In November 1986, *The Times* newspaper published reports concerning pollution of the River Rhine. One report is reproduced on the right. Read the report on the right, and the article below.

Safety standards questioned

The Sandoz chemical plant in Basle has been accused of insufficient safety standards after a fire which led to serious pollution of the Rhine.

A report suggests that the storage building where the fire broke out had no trays to catch leaking chemicals, no automatic sprinklers and no smoke-warning or heat warning systems.

The building stored a powerful fungicide containing mercury. Only 10 metres from this building, there was another store containing sodium. If water, which was used to fight the fire, had reacted with the sodium a serious explosion might have resulted.

An investigation into the cause of the fire will probably take weeks to prepare. Meanwhile, many German towns along the Rhine face a water shortage because their wells have been closed.

In Basle, demonstrators marched through the city to the Sandoz Chemical Company. They carried banners saying, ' We don't want to be tomorrow's fish.'

The demonstrators were angry about reports that at least 200 gallons of a poisonous liquid containing mercury had leaked into the River Rhine. The West Geman government may demand damages for the accident and press for more co-ordination between countries over inland waterway disasters.

The Rhine Fishing Federation claimed that plants and animals would be killed in 155 miles of the Rhine between Basle and Manheim. Salvage workers had recently removed several hundred kilograms of dead eels from the river.

The enquiry

Imagine that an enquiry has been set up to look into the accident. Some people have been asked to prepare statements for the enquiry team. These people include:
(i) a scientist who is particularly interested in the toxic (poisonous) properties of mercury and its compounds,
(ii) a safety officer required to enforce safety standards in industrial plants,
(iii) the Mayor of Basle.

The statements

Make a group with two other students. Decide who will act as the scientist, who will act as the safety officer and who will act as the Mayor of Basle. Discuss the issues in your group. Then, write down the statements you will each make to the enquiry team (about 300 words). Practise your talks to each other. (Your teacher may ask you to read your statements to the rest of the class.)

Pollution alert in North Sea
By Pearce Wright

Ships which monitor North Sea pollution and radioactivity levels for the Ministry of Agriculture and Fisheries have been alerted to track a flood of mercury due to be discharged from the Rhine later this week.

Their measurement of how this lethal plume is dispersing will be relayed to experts on the protection of coastal and offshore North Sea fisheries.

Aquatic life in the Rhine was destroyed as an estimated 30 tons of mercury and other chemicals

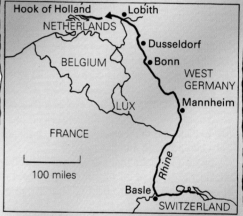

used in the manufacture of pesticides drifted down the river from Basle in Switzerland, through Germany, and to the Netherlands at the weekend.

The mercury, the key ingredient of a powerful fungicide, was washed into the river when firemen fought a blaze at the Sandoz chemical plant, near Basle.

It undid efforts of 10 years to clean up the Rhine, which had been criticized as "becoming Europe's sewer" because of the effluent from rapid industrial growth in Germany, Switzerland and France.

Mercury is one of the heavy metals that are mined for various industrial purposes. Others include lead, arsenic and cadmium. They are also called the toxic metals because very small concentrations are required to poison animals and plants.

A similar incident with mercury 25 years ago, though spread over a longer period, gave the first hints of the extreme toxicity of the heavy metals.

It happened in the small Japanese coastal town of Minamata, where mercury poisoning originating in wastes from a chemical factory spread from fish to fishermen and their families.

The disaster occurred because the mercury was transformed into a highly biologically active form of organic mercury compound after it was discharged. That anxiety will exist over the mercury pouring into the North Sea.

Section B: Study Questions

1 Many fuels contain small amounts of sulphur. This forms sulphur dioxide when the fuel is burned. The sulphur concentration in lichens was measured at various points in a town. The map shows points (A–F) where measurements were made. The table below shows the results.

Point	Sulphur concentration/ parts per million
A	0.02
B	3.30
C	2.10
D	1.50
E	0.01
F	0.02

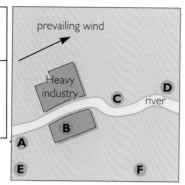

(a) Write an equation to show the formation of sulphur dioxide when a fuel is burnt.
(b) Why did the lichens contain traces of sulphur?
(c) How does sulphur dioxide cause acid rain?
(d) Study the map and the results. What conclusions do you make?
(e) Explain the difference in sulphur concentration between points A and C.
(f) Since 1955, the number of species of flowering plants within 1 kilometre of the town centre was monitored, together with the output of sulphur dioxide.

Year	Millions of tonnes of sulphur dioxide	Numbers of species of flowering plants
1955	5.6	442
1965	6.5	330
1970	5.8	322
1975	5.4	320
1980	5.4	317
1985	5.1	317

(i) How does the sulphur dioxide seem to be affecting the flowering plants?
(ii) Why do you think the number of species of flowering plants has fallen?
(iii) What steps should be taken to enable the number of species of flowering plants to increase?

2 The apparatus, above right, can be used to find the composition of the air. 100 cm³ of air were placed in syringe A with syringe B empty. The copper was heated strongly and the air was passed to and fro between syringes A and B over the hot copper, and finally returned to syringe A.

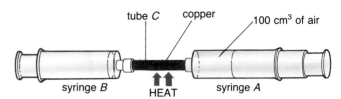

(a) What compound is formed in tube C?
(b) Which gas does the copper remove from the air?
(c) What volume of gas is left in syringe A at the end of the experiment?
(d) What is the main gas remaining in syringe A at the end of the experiment?

3 The Clean Air Act of 1956 made the emission of dark smoke from a chimney illegal. A definition of dark smoke was made in terms of a white card with a scale composed of areas with differing degrees of black shading. New industrial plants burning solid fuel at rates of 10 tonnes per hour or more must install efficient devices to trap the grit and dust produced.
(a) How would you construct a shaded card to use for testing the darkness of smoke emitted from a chimney?
(b) How could you represent the extent of darkness of the smoke using this shaded card?
(c) What devices could industrial plants use to trap grit and dust?
(d) How would you design and locate (i) industrial furnaces, (ii) industrial chimneys, to reduce smoke pollution?

4 (a) What processes cause smoke pollution in the atmosphere?
(b) Why do industrial plants get rid of their smoke through tall chimneys?
(c) What effect does smoke pollution of the atmosphere have on health? Which diseases are most linked to smoke pollution?
(d) Devise experiments (i) to show that city air is polluted with smoke particles, (ii) to obtain clean air.

5 Since the year 1900, the percentages of carbon dioxide and lead compounds in the atmosphere have increased. Suggest reasons for the increase in (i) carbon dioxide, (ii) lead compounds.

6 Steam is passed over red-hot coke and a mixture of two gases (A and B) is produced. This mixture is then burned in pure oxygen and, when the products of the reaction are cooled to room temperature, a colourless liquid, C, and a colourless gas, D, remain. D reacts with a dilute solution of calcium hydroxide (lime water) to form a white precipitate, E. D allows magnesium to burn in it, forming two solids, F and G. When D is passed over red-hot coke, the gas A is formed.
(a) Identify the substances A–G.
(b) Write word equations for all the reactions involved.

SECTION C
Particles

A painting of John Dalton collecting marsh gas which is mainly methane. John Dalton was the first scientist to use the word 'atom'

1 Evidence for Particles

What happens to ice when it melts? How does the ice manage to change into water?

When sugar is added to tea or coffee, it dissolves quite quickly. The sugar seems to disappear. What happens to the sugar when it dissolves? Where does the sugar go?

How does the liquid get through the filter paper when filter coffee is made? The coffee grains don't get through the paper. Why is this?

Look at the photos above and try to answer the questions in the captions.

In order to answer these questions you will need to use the idea that

> All materials are made up of particles.

If possible, get some sugar and stir it into warm water. Watch the sugar disappear as you stir the water. We can explain how the sugar dissolves and disappears using the idea of particles. Both sugar and water are made up of very small particles. The particles are much too small to see, even under a microscope. When sugar dissolves, tiny particles break off each solid granule. These sugar particles mix with the water particles in the liquid. So, the solution can taste sweet even though you cannot see the sugar (figure 1).

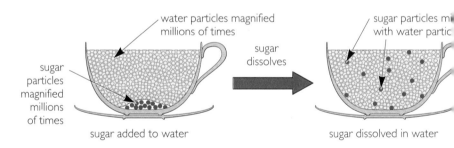

Figure 1
Explaining what happens when sugar dissolves in water

Only a small amount of curry powder gave this meal its hot taste

How large are the particles of materials?

Anyone who cooks knows that a small amount of pepper, ginger or curry powder will give something a really strong taste. Too much spice can spoil the whole meal. This suggests that tiny particles in the spice can spread throughout the whole dish. Figure 2 shows an experiment which will help you to get some idea about the size of particles.

Dissolve I g of dark purple potassium manganate(VII) crystals in 1000 cm^3 of water. Take 100 cm^3 of this solution and dilute it to 1000 cm^3 with water. Now take 100 cm^3 of the once-diluted solution and dilute this to 1000 cm^3 with water. Repeat the dilution again and again until you get a solution in which you can only *just* see the pink colour. It is possible to make 6 dilutions before the pink colour is so faint that it is only just noticeable. When the potassium manganate (VII) dissolves, its particles spread throughout the water making a dark purple solution. When this solution is diluted, the particles spread further apart.

This experiment shows that the tiny particles in only I g of potassium manganate(VII) can colour I 000 000 000 cm^3 of water. This suggests that there must be millions and millions of tiny particles in only I g of potassium manganate(VII).

Similar experiments show that the particles in all substances are extremely small. For example, there are more air particles in a thimble than grains of sand on a large beach.

When oil is spilt on water or on the road, it covers the surface in a thin, brightly-coloured film

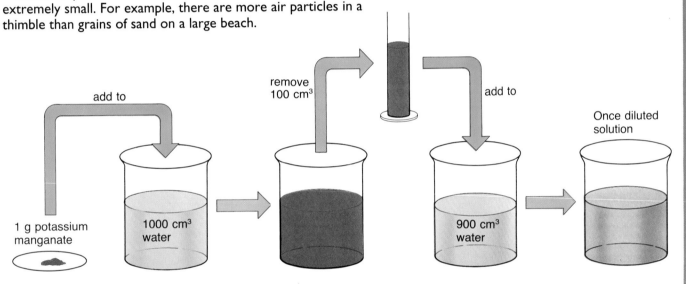

add to

remove 100 cm^3

add to

Once diluted solution

1 g potassium manganate

1000 cm^3 water

900 cm^3 water

Figure 2 If you try this experiment wear eye protection

Questions

I Krisnan and Christine were talking about dissolving sugar in tea. Krisnan thought that the sugar would weigh less when it was dissolved because it would be floating in the tea.
(a) What do you think happens to the mass of a substance when it dissolves?
(b) Plan an experiment to test your suggestions in part (a).
(c) If possible, carry out your suggested experiment. Explain your results.

2 Get into groups of two or three. Use the idea of particles to discuss and explain what happens when:
(a) water in a kettle boils to produce steam,

(b) you add water to a clay flower pot and the outside of the pot becomes wet,
(c) puddles disappear on a fine day,
(d) tightly tied balloons go down after some time.

3 Look back at figure 2 and the experiment involved. Suppose that I g of potassium manganate(VII) has a volume of I cm^3 and that there is one particle of potassium manganate(VII) in every drop of the final I 000 000 000 cm^3 of faint pink solution.
(a) Estimate the number of drops in I cm^3 of the faint pink solution.
(b) How many particles of potassium manganate(VII) are there

in I 000 000 000 cm^3 of the faint pink solution?
(c) Calculate the volume of one particle of potassium manganate(VII) in the crystal.

4 A goldsmith used 1.93 g of gold (density 19.3 g/cm^3) to make an extremely thin sheet of gold, 100 cm^2 in area.
(a) What is the volume of the gold foil? (Use the density to obtain this.)
(b) What is the thickness of the gold foil?
(c) What is the largest possible size of gold particles?

2 Particles in Motion

Why is it possible to smell the perfume that someone is wearing from several metres away?

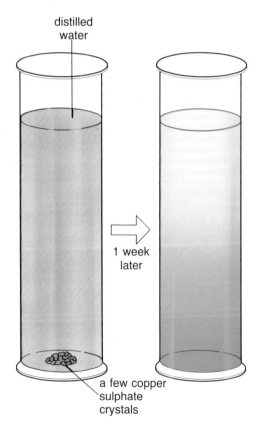

Figure 2
Why is the blue colour darker near the bottom of the gas jar? Why will all the solution eventually be the same colour?

The best evidence that particles of matter are constantly moving comes from studies of *diffusion* and *Brownian motion*.

- **Diffusion.** Fish and chips have a delicious smell. How does the smell get from the fish and chips to your nose?

  Particles of gas are released from the fish and chips. These particles mix with air particles and move away from the chips. *This movement and mixing of particles is called* **diffusion**.

  Gases diffuse to fill all the space available to them—even heavy gases like bromine (figure 1). How does the bromine vapour get to the top of the gas jar? Scientists believe that gases consist of tiny particles, moving around haphazardly and colliding with each other and the walls of their container. The gas particles don't care where they go, so sooner or later they will spread into all the space available.

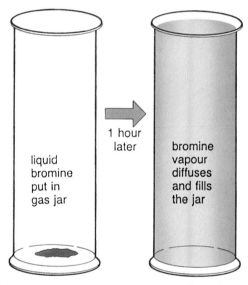

Figure 1 Liquid bromine evaporates and the vapour diffuses Care: Bromine is very toxic (poisonous). You should never handle it yourself

Diffusion also occurs in liquids, although it takes place more slowly than in gases (figure 2). This means that liquid particles move around more slowly than gas particles. Diffusion does not happen in solids.

Diffusion is very important in living things. It explains how the food you eat gets to different parts of your body. After a meal, food passes into the stomach and through the intestines. Large particles are broken down into smaller ones. The smaller particles can then diffuse through the walls of the intestines into the bloodstream.

- **Brownian motion.** In 1827 a biologist called Robert Brown was using a microscope to look at pollen grains in water. To his surprise, the pollen grains kept moving and jittering about randomly. Similar random movements can be seen when you look at smoke particles through a microscope (figure 3). *This movement of tiny particles in a gas or liquid is called* **Brownian motion**.

  Smoke from a smouldering piece of string is injected into the smoke cell using a teat pipette. Under the microscope, the smoke particles look like tiny pinpoints of light which jitter about (figure 3).

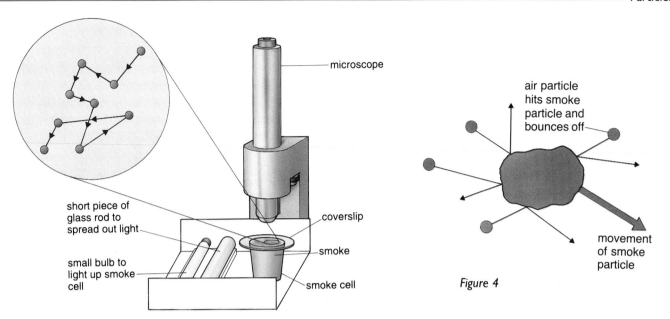

short piece of
glass rod to
spread out light

small bulb to
light up smoke
cell

microscope

coverslip

smoke

smoke cell

Figure 3

air particle
hits smoke
particle and
bounces off

movement
of smoke
particle

Figure 4

The movement of the smoke particles is caused by the random motion of oxygen and nitrogen particles in the air around them. The particles of smoke are small, but they are much larger than air particles. Through the microscope we can see smoke particles, but air particles are much too small to be seen. These air particles move very fast and hit the smoke particles at random. The smoke particles are therefore knocked first this way and then that way so they appear to jitter about (figure 4). As the temperature rises, particles have more energy and they move about faster. This means that gases and liquids diffuse faster when the temperature rises. Particles undergoing Brownian motion also jitter about faster as the temperature rises because they are being bumped more often by the small particles in the gas or the liquid.

Gas pressure

Gas pressure is caused by moving particles. It helps to inflate balloons and tyres (figure 5).

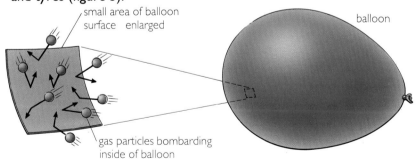

small area of balloon
surface   enlarged

balloon

gas particles bombarding
inside of balloon

Figure 5
Gas particles move randomly inside the balloon. They collide with each other and with the walls of the balloon. Constant bombardment of the inside of the balloon has a gas pressure on its outside and its inside. The gas pressure inside is greater than the gas pressure outside. The extra pressure inside the balloon keeps it inflated

Questions

1 What is (i) diffusion, (ii) Brownian motion?
2 (a) Why is it important to switch off the engine when filling a car's petrol tank?
(b) Why is it possible to smell some cheeses even when they are wrapped in cling film?
3 You can smell hot, sizzling bacon several metres away, yet you have to be near cold bacon to smell it. Explain this difference in terms of particles.
4 (a) Why does bromine vapour eventually fill the gas jar in figure 1?
(b) How would the bromine diffuse if all the air was removed from the gas jar?

3 Particles in Mixtures

When salt is mixed with water, a clear **solution** is produced. This clear solution is a **mixture**. Tiny particles of salt are spread evenly throughout water in the solution. It is impossible to see the particles of salt in the solution because they are so small.

When mud is mixed with water, a cloudy mixture is produced. If the water is decanted from its sediment of mud, it still looks cloudy. In this case, particles of mud remain 'floating' in the water. The mud particles are much larger than the salt particles in salt solution. The mud particles cannot dissolve in the water, so a cloudy liquid is formed. This cloudy liquid is an example of a **colloid**. A colloid is made from two substances which cannot dissolve in each other or mix properly with each other. It contains bits or globules or particles of one substance spread through the other. These particles are too large to dissolve, but too small to settle out.

The photographs below show some examples of important colloids.

*Most dairy products are colloids. They consist of tiny drops of liquid spread throughout another liquid or a solid. These are called **emulsions**. Butter and cheese contain tiny globules of water spread throughout a fatty solid. Milk and yoghurts contain tiny globules of fat spread throughout a watery liquid*

*Mist, fog and clouds consist of tiny droplets of water suspended in the air. Colloids with tiny particles of solid or liquid suspended in a gas are called **aerosols***

*Shaving cream and soap suds consist of small bubbles of air spread throughout soap. Colloids with tiny bubbles of gas in a solid or liquid are called **foams**. Expanded polystyrene is an example of a solid foam*

Emulsion paint is another important example of a colloid. This contains three different kinds of particles suspended in water. These are particles of a white pigment (titanium oxide), particles of latex and particles of china clay.

The particles in emulsion paint are not visible to the naked eye

In colloids, the particles, droplets or globules of one material are mixed up with the other, but not dissolved in it. The particles are scattered about or *dispersed* in a random manner. In fact, every colloid has two parts. One part is the *continuous* material like water in milk or air in mist. The other part is the *dispersed* material like fat globules in milk or water droplets in mist (Figure 1).

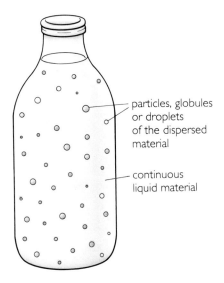

particles, globules
or droplets
of the dispersed
material

continuous
liquid material

Figure 1
The continuous and dispersed materials in a colloid

Questions

1 There are eight different types of colloid. These are listed in the table below.

Continuous material	Dispersed material	Type of colloid
gas	liquid	aerosol
gas	solid	aerosol
liquid	gas	foam
liquid	liquid	emulsion
liquid	solid	sol
solid	gas	solid foam
solid	liquid	solid emulsion
solid	solid	solid sol

(a) Emulsion paint is not really an emulsion. What is it?
(b) Why is it not possible to make a colloid from two gases?
(c) What type of colloid are the following?
pumice, ice cream, bread, whipped cream, smoke, deodorant spray, currant cake, expanded polystyrene, lipstick, suntan lotion, gloss paint, clouds.

2 Jelly is made by mixing gelatine with warm water. Gelatine is a protein which consists of very large molecules. The gelatine disperses in the water. It does not dissolve. As the mixture of gelatine in water cools, a **gel** is formed. In this gel, water droplets are trapped in a continuous network of gelatine.

(a) What type of colloid is
(i) gelatine in warm water, (ii) the jelly when it has set?
(b) Draw a diagram to illustrate the colloid of gelatine in warm water.
(c) Draw a diagram to illustrate the colloid in the jelly when it has set.

3 (a) Measure the diameter of the solid particles in the emulsion paint in the photograph above.
(b) Use the magnification given in the caption to calculate the *actual* diameter of the particles.
(c) How many times are these particles larger than single atoms? (The diameter of an atom is about $\frac{1}{100\,000\,000}$ cm $= 10^{-8}$ cm.)

4 Change of State

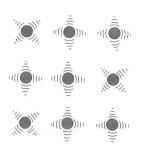

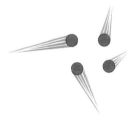

Figure 1

Figure 2

Figure 3

The kinetic theory of matter

The idea that matter is made of moving particles is called the **kinetic theory**. (The word 'kinetic' comes from a Greek word *kineo* which means 'I move'.) The main points of the theory are as follows:

1. All matter is made up of tiny, invisible, moving particles.
2. Particles of different substances have different sizes.
3. Small particles move faster than heavier ones at a given temperature.
4. As the temperature rises, the particles move faster.
5. In a solid, the particles are very close and they can only vibrate about fixed positions (figure 1).
6. In a liquid, the particles are a little further apart. They have more energy and they can move around each other (figure 2).
7. In a gas, the particles are far apart. They move rapidly and randomly in all the space they can find (figure 3).

The movement of particles in solids, liquids and gases can be compared to the movement of children in school. During lessons, the pupils stay in their seats but they are not still. They write notes, answer questions and wriggle (*vibrate*) about in their seats. They are behaving like the particles of a *solid*. When the lesson ends, the pupils pack up their things and move in a group to another part of the school for the next lesson. The pupils stay together, but *move around each other*. They resemble the particles in a *liquid*. When the bell rings at break, the children spill out into the playground where they *move around freely and randomly*. The pupils now resemble a *gas*.

During lessons pupils resemble solid particles

Between lessons pupils resemble liquid particles

In the playground pupils resemble gas particles

Changes of state

Solids, liquids and gases are sometimes called the three **states of matter**. The kinetic theory can be used to explain how a substance changes from one state to another. A summary of the different changes of state is shown in figure 4. These changes are usually caused by heating or cooling a substance.

- **Melting and freezing.** When a solid is heated, the particles gain energy and vibrate faster and faster. Eventually, they break free from their fixed position and begin to move round each other. The solid melts to form a liquid. *The temperature at which the solid melts is the* **melting point**.

The temperature at which a solid melts tells us how strongly its particles are held together. Substances with high melting points have strong forces between their particles; substances with low melting points have weak forces between their particles. Metals and alloys, like iron and steel, have high melting points. This suggests that there are strong forces between their particles. This is why metals can be used as girders and supports.

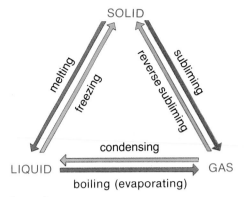

Figure 4

The forces between metal particles in steel are so strong that steel can be used in thin cables to lift heavy loads

- **Evaporating and boiling.** When a liquid is heated, the particles gain energy and move around each other faster and faster. Some particles near the surface of the liquid have enough energy to escape from those around them in the air, and some of the liquid evaporates to form a gas.

Eventually, a temperature is reached at which the particles are trying to escape from the liquid so rapidly that bubbles of gas actually start to form inside the liquid. *The temperature at which this evaporation begins to occur within the bulk of the liquid is the* **boiling point**. Liquids which evaporate at low temperatures are described as *volatile*.

The temperature at which a liquid boils tells us how strongly the particles are held together in the liquid. Liquids with high boiling points have stronger forces between their particles than liquids with low boiling points.

Gases fill their container completely

Questions

1 How do the particles move in (i) a solid; (ii) a liquid; (iii) a gas?

2 What do you understand by the following:

kinetic theory; states of matter; melting point; boiling point?

3 What happens to the particles of a liquid (i) as it cools down; (ii) as it freezes?

4 Use the kinetic theory to explain why (i) gases exert a pressure on the walls of their container; (ii) solids have a fixed size and a fixed shape; (iii) liquids have a fixed size but not a fixed shape; (iv) solid blocks of air freshener used in toilets can disappear without leaving any solid.

5 Atoms—The Smallest Particles

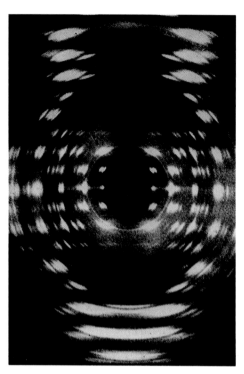

This X-ray photograph of DNA has bright spots caused by X-rays being reflected by particles in the crystal

Dalton's theory of atoms

Experiments involving diffusion and Brownian motion (unit 2 of this section) show that substances are made up of small particles. Photographs taken by X-rays and electron microscopes also provide good evidence for particles. X-ray photographs of crystals, such as the one on the left, show white spots on a black background. The white spots are caused by X-rays being reflected by regularly-spaced particles in the crystal.

Electron microscopes can magnify objects more than a million times. In 1958, scientists in the USSR observed individual particles of barium and oxygen using an electron microscope. Figure 1 shows an electron microscope photo of a manganese compound. The magnification is 127 000 times. The dark lines are rows of manganese particles in the crystal. These particles are the smallest possible particles of manganese—they are manganese *atoms*.

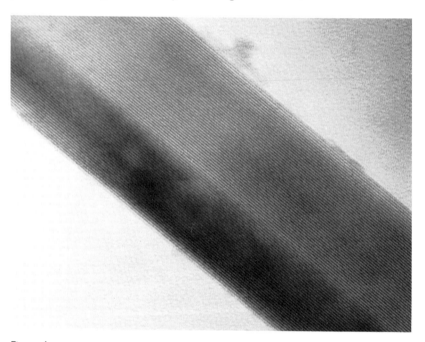

Figure 1
This photograph of a manganese compound was taken through an electron microscope. The dark lines are rows of manganese atoms

The smallest particle of any element is called an **atom**. The word 'atom' was first used by John Dalton in 1807, when he put forward his **Atomic Theory of Matter**. The main points in Dalton's theory are:

1 All matter is made up of tiny particles called atoms.
2 Atoms cannot be made or destroyed.
3 Atoms of the same element are alike, with the same mass, colour, etc.
4 Atoms of different elements have different masses, colours, etc.
5 Atoms can join together to form larger particles in compounds.

Nowadays, we call these larger particles **molecules**. Dalton's theory about atoms and molecules is still very useful.

John Dalton—the first scientist to use the name 'atom' for the smallest particle of an element. Dalton was born in 1766 in the village of Eaglesfield in Cumbria. His father was a weaver. For most of his life, Dalton taught at the Presbyterian College in Manchester

Representing atoms with symbols

Dalton also suggested a method of representing atoms with symbols. Figure 2 shows how he represented the atoms of some common elements. The modern symbols which we use for different elements are based on his suggestions.

The table gives a list of the symbols of some of the common elements. (A longer list of symbols appears on page 283.) Notice that most elements have two letters in their symbol; the first letter is a capital, the second letter is *always* small. These symbols come from either the English name (O for oxygen, C for carbon) or the Latin name (Au for gold—Latin: *aurum*; and Cu for copper—Latin: *cuprum*).

Element	Symbol	Element	Symbol	Element	Symbol
Aluminium	Al	Helium	He	Oxygen	O
Argon	Ar	Hydrogen	H	Phosphorus	P
Bromine	Br	Iodine	I	Potassium	K
Calcium	Ca	Iron	Fe	Silicon	Si
Carbon	C	Lead	Pb	Silver	Ag
Chlorine	Cl	Magnesium	Mg	Sodium	Na
Chromium	Cr	Mercury	Hg	Sulphur	S
Cobalt	Co	Neon	Ne	Tin	Sn
Copper	Cu	Nickel	Ni	Uranium	U
Gold	Au	Nitrogen	N	Zinc	Zn

The symbols for some elements

Using symbols, we can represent compounds as well as elements. For example, water is represented as H_2O because the smallest particle of water (a molecule) contains two hydrogen atoms and one oxygen atom. Carbon dioxide is written as CO_2—one carbon and two oxygen atoms. H_2O and CO_2 are called **formulas**. Formulas show the relative numbers of atoms of each element in a compound.

Chemists use symbols in the same way that typists use shorthand. Using symbols, they can represent substances and chemical changes quickly. For example, the formula for potassium manganate (VII) is $KMnO_4$. If you have to name this substance many times, how would you prefer to write it—using symbols or using the long name!

If a substance is pure it will contain only one kind of material. So, if a coloured substance is pure it will contain only one ring or band during chromatography

By using shorthand you can make notes more quickly. Chemists use chemical symbols as their shorthand

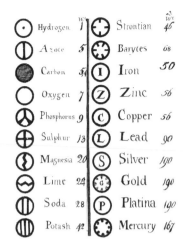

Figure 2
Dalton's symbols for some common elements

Questions

1 Look at the electron microscope photograph of a manganese compound, on the opposite page.
   (a) Estimate the thickness of one of the dark lines in the photo. This dark line is a row of manganese atoms.
   (b) Calculate the actual diameter of a manganese atom. (Assume the magnification is 127 000.)
2 Find out about the life and work of John Dalton. Prepare a short talk about Dalton for the rest of your class. If possible, illustrate your talk with drawings, diagrams and photos.
3 How many atoms of the different elements are there in one molecule of (i) methane (natural gas), CH_4, (ii) sulphuric acid, H_2SO_4, (iii) sugar, $C_{12}H_{22}O_{11}$, (iv) chloroform, $CHCl_3$?
4 Look at Dalton's symbols for the elements in Figure 2.
   (a) What do we call 'Azote' and 'Platina' today?
   (b) Six of the substances in Dalton's list are compounds and *not* elements. Pick out two of these compounds and write their correct chemical names.
   (c) Which one of Dalton's symbols do you think is the most appropriate? Why do you think it is appropriate?
   (d) Look at the numbers at the right-hand side of the elements in Dalton's list. What do you think these numbers represent?

6 Measuring Atoms

How large are atoms?

Experiments with thin films of oil on water show that olive oil particles are about $\frac{1}{100\,000\,000}$ cm thick. But olive oil particles are large molecules containing more than 50 atoms. If we estimate that one molecule of olive oil is about 10 atoms thick, how big is a single atom?

Atoms are about $\frac{1}{100\,000\,000}$ of a centimetre across. This means that if you put 100 million of them side-by-side, they will measure only 1 cm. It is very difficult to imagine anything as small as this. You will get some idea of the size of atoms from figure 1. If atoms were magnified to the size of marbles then, on the same scale, marbles would have a diameter of 1500 km—one third of the distance between New York and London.

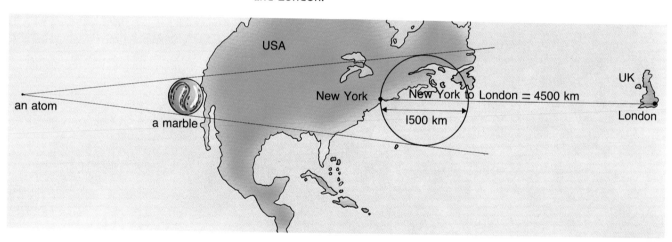

Figure 1
If atoms are magnified to the size of marbles, on the same scale a marble would have a diameter of 1500 kilometres

Figure 2 will also help you to realize just how small atoms are. It shows a step-by-step decrease in size from 1 cm to $\frac{1}{10\,000\,000}$ cm. Each object is one hundred times smaller than the one before it. The dice on the left is about 1 cm wide. In the next picture, the grain of sand is about $\frac{1}{100}$ cm across. The bacterium in the middle is 100 times smaller again—about $\frac{1}{10\,000}$ (10^{-4}) cm from end to end. In the next picture, the molecule of haemoglobin is 100 times smaller than this—about $\frac{1}{1\,000\,000}$ (10^{-6}) cm in diameter. Finally, in the right-hand picture, the atom is one hundredth of the size of the haemoglobin molecule— about $\frac{1}{100\,000\,000}$ (10^{-8}) cm in diameter.

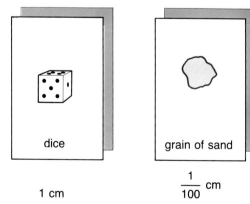
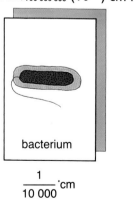
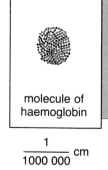
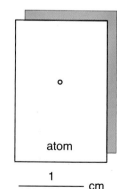

dice	grain of sand	bacterium	molecule of haemoglobin	atom
1 cm	$\frac{1}{100}$ cm	$\frac{1}{10\,000}$ cm	$\frac{1}{1\,000\,000}$ cm	$\frac{1}{100\,000\,000}$ cm

Figure 2
Step-by-step to the size of atoms

How heavy are atoms?

A single atom is so small that it cannot be weighed on a balance. However, *the mass of one atom can be compared with that of another atom using an instrument called* **a mass spectrometer** (figure 3).

In a mass spectrometer, atoms are passed along a tube and focused into a thin beam. This beam of particles passes through an electric field (which speeds them up) and then through a magnetic field where they are deflected. The extent to which an atom is deflected depends on its mass—the greater the mass, the smaller the deflection. Using the amount of deflection, it is possible to compare the masses of different atoms and make a list of their relative masses.

Element	Symbol	Relative atomic mass
Carbon	C	12.0
Hydrogen	H	1.0
Oxygen	O	16.0
Iron	Fe	55.8
Copper	Cu	63.5
Gold	Au	197.0

Table 1: The relative atomic masses of a few elements

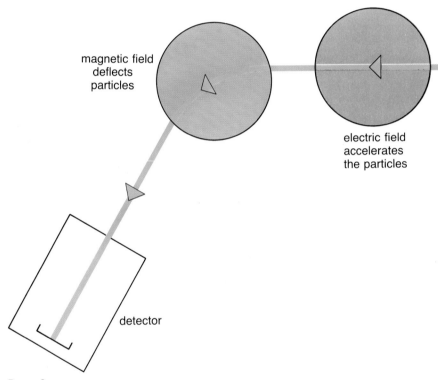

magnetic field deflects particles

electric field accelerates the particles

parallel slits to produce a thin beam

beam of atoms

detector

Figure 3
A simple diagram of a mass spectrometer

Relative atomic masses

Chemists compare the masses of different atoms using **relative atomic masses** (RAMs). The symbol A_r is sometimes used for relative atomic mass. The element carbon is chosen as the standard against which the masses of other atoms are compared. Carbon atoms are given a relative atomic mass of 12 and the relative masses of other atoms are obtained by comparison with this. A few relative atomic masses are listed in Table 1. Other relative atomic masses are given on page 283. From table 1 you will see that carbon atoms are 12 times as heavy as hydrogen atoms and iron atoms are 55.8 times as heavy as hydrogen atoms.

Questions

1 The radius of a potassium atom is $\frac{2}{100\,000\,000}$ cm. How many potassium atoms can be arranged next to each other to make a line 1 cm long?

2 Use the relative atomic masses on page 283 to answer the following questions.
   (a) Which element has the lightest atoms?
   (b) Which element has the next lightest atoms?
   (c) How many times heavier are carbon atoms than hydrogen atoms?
   (d) Which element has atoms four times as heavy as oxygen?

3 Put the following in order of size from the largest to the smallest:
a bacterium; the thickness of a human hair; a molecule of sugar (which contains about 50 atoms); a smoke particle; a copper atom; a fine dust particle.

4 Write the following elements in order of decreasing deflection in a mass spectrometer (put the element which is deflected the most first):
copper; calcium; carbon; cobalt; chlorine.

7 Using Relative Atomic Masses

The photograph shows an aluminium colander (mass 108 g), a copper bracelet (mass 16 g), some iron nails (mass 5.58 g) and some barbecue charcoal (mass 150 g). How many moles of each element do the objects contain? (Al = 27, Cu = 64, Fe = 55.8, C = 12)

Chemists are not the only people who 'count' by weighing. Bank clerks, like the one in the photo, use the same idea when they count coins by weighing them. For example, one hundred 1p coins weigh 356 g; so it is quicker to take one hundred 1p coins by weighing 356 g of them than by counting

Relative atomic masses show that one atom of carbon is 12 times as heavy as one atom of hydrogen. Therefore, 12 g of carbon will contain the same number of atoms as 1 g of hydrogen. An atom of oxygen is 16 times as heavy as an atom of hydrogen, so 16 g of oxygen will also contain the same number of atoms as 1 g of hydrogen. In fact, *the relative atomic mass (in grams) of every element* (1 g of hydrogen, 12 g carbon, 16 g oxygen, etc.) *will contain the same number of atoms*. This number is called **Avogadro's constant** in honour of the Italian scientist Amedeo Avogadro. *The relative atomic mass in grams is known as one* **mole** *of the element.* So, 12 g of carbon is 1 mole of carbon and 1 g of hydrogen is also 1 mole. 24 g of carbon is 2 moles and 240 g of carbon is 20 moles.

$$\text{Notice that the number of moles} = \frac{\text{mass}}{\text{RAM}}$$

Experiments show that Avogadro's constant is 6×10^{23}. Written out in full this is 600 000 000 000 000 000 000 000. Thus *1 mole of an element always contains 6×10^{23} atoms.*

We can use the mole idea to count (calculate) the number of atoms in a sample of element. For example:

12 g (1 mole) of carbon contains 6×10^{23} atoms
so 1 g ($1/12$ mole) of carbon contains $1/12 \times 6 \times 10^{23}$ atoms
$\Rightarrow$ 10 g ($10/12$ mole) of carbon contains $10/12 \times 6 \times 10^{23} = 5 \times 10^{23}$ atoms
i.e. number of atoms = number of moles $\times 6 \times 10^{23}$

$$\text{i.e. number of atoms} = \text{number of moles} \times 6 \times 10^{23}$$

Chemists often need to count atoms. In industry, nitrogen is reacted with hydrogen to form ammonia, NH_3, which is then used to make fertilizers. In a molecule of ammonia, there is one nitrogen atom and three hydrogen atoms. In order to make ammonia, chemists must therefore react:

1 mole of nitrogen + 3 moles of hydrogen
(14 g of nitrogen) (3 × 1 g = 3 g of hydrogen)

not 1 g of nitrogen and 3 g of hydrogen.

To get the right quantities, chemists must measure in moles *not* in grams. Thus the mole is the chemist's counting unit.

Finding formulas

We have used some formulas already, but how are they obtained? How do we know that the formula of water is H_2O?

All formulas are obtained by doing experiments to find the masses of elements which react. When water is decomposed into hydrogen and oxygen, results show that

18 g of water give 2 g of hydrogen + 16 g of oxygen
= 2 moles of hydrogen + 1 mole of oxygen
= $2 \times 6 \times 10^{23}$ atoms + 6×10^{23} atoms
of hydrogen   of oxygen

Since 12×10^{23} hydrogen atoms combine with 6×10^{23} atoms of oxygen, it means that 2 hydrogen atoms combine with 1 oxygen atom. Therefore, the formula must be H_2O. These results are set out in table 1. By finding the masses of reacting elements, we can use relative atomic masses to calculate the number of moles of atoms that are present and this gives us the formula.

	H	O
Masses present	2 g	16 g
Mass of 1 mole	1 g	16 g
∴ moles present	2	1
Ratio of atoms	2	1
⇒ Formula	H_2O	

Table 1: Finding the formula of water

Finding the formula of magnesium oxide

When magnesium ribbon is heated, it burns with a very bright flame to form white, powdery magnesium oxide:

$$\text{magnesium} + \text{oxygen} \rightarrow \text{magnesium oxide}$$

If you try this experiment **wear eye protection**.
Weigh accurately 0.24 g of clean magnesium ribbon. Heat this strongly in a crucible until all of it forms magnesium oxide (figure 1). Put a lid on the crucible to stop magnesium oxide escaping, but keep a small gap so that air can enter. When the magnesium has finished reacting, reweigh the crucible + lid + magnesium oxide.

Calculate the mass of oxygen reacting from the mass of magnesium oxide minus the mass of magnesium.

Table 2 shows how to obtain the formula of magnesium oxide from the results.

Mass of magnesium reacting		= 0.24 g.
Mass of magnesium oxide produced		= 0.40 g.
	Mg	**O**
Masses reacting	0.24 g	0.16 g
Mass of 1 mole	24 g	16 g
∴ moles present	0.01	0.01
Ratio of moles	1	1
∴ ratio of atoms	1	1
⇒ Formula	MgO	

Table 2: Finding the formula of magnesium oxide

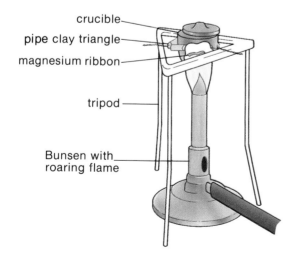

crucible
pipe clay triangle
magnesium ribbon
tripod
Bunsen with roaring flame

Figure 1

Questions

1 (a) What is the relative atomic mass of hydrogen?
(b) How many atoms are there in 1 g of hydrogen?
(c) How many atoms are there in 20 moles of hydrogen atoms?
(d) How heavy are 20 moles of hydrogen atoms?
(e) How many moles are 24×10^{23} hydrogen atoms?

2 How many atoms are there in (i) 52 g chromium (Cr = 52) (ii) 2.8 g nitrogen (N = 14) (iii) 0.36 g carbon (C = 12) (iv) 20 g bromine (Br = 80)?

3 What is the mass of (i) 3 moles of bromine (Br = 80) (ii) ¼ mole of calcium (Ca = 40) (iii) 0.1 mole of sodium (Na = 23)?

4 Methane in natural gas is found to contain 75% carbon and 25% hydrogen. Calculate the formula of methane using a method like that in table 1.

5 What are the formulas of the following compounds?
(a) A compound in which 10.4 g chromium (Cr = 52) combines with 48 g bromine (Br = 80).
(b) A nitride of chromium in which 0.26 g chromium forms 0.33 g of chromium nitride (N = 14).

8 Particles in Elements

Look at the properties of iron and oxygen in Table 1. How do the properties of metals, like iron differ from those of non-metals like oxygen? We can use our ideas about particles in elements and in liquids, solids and gases to explain the properties of iron and oxygen. This will also help you to understand the properties of metals and non-metals.

Property	Iron	Oxygen
State at room temperature	solid	gas
Melting point	high	very low
Boiling point	high	very low
Density	high	very low
Electrical conductivity	good	poor
Squashiness (compressibility)	very hard	easy to squash

Table 1: The properties of oxygen and iron

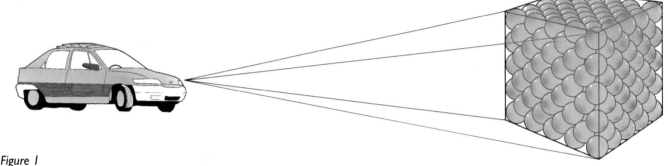

Figure 1
In iron and most other metals, single atoms are packed close together and in a regular pattern

large region containing tiny negative particles called electrons

small positive central core called a nucleus

Figure 2
The structure of an atom

Iron and metals

In iron and other metals, the particles are single atoms. These atoms are packed together very closely and in a regular fashion (figure 1). The regular arrangement of particles builds up to give a solid with a regular shape called a **crystal**. The atoms are packed so tightly that there are strong forces between them. This gives the metal a high density and makes it hard and strong. The strong forces hold the atoms together very tightly and restrict their movement. This results in a high melting point and high boiling point.

All atoms have a small positively-charged core or **nucleus** surrounded by tiny negative particles called **electrons** (figure 2). In metals, the electrons are fairly free to move around the nucleus and away from it. They can easily be attracted to the positive terminal of a battery and form an electric current. This is why metals conduct electricity. In non-metals the electrons cannot move so easily. So non-metals are usually poor conductors.

The strong forces between metal atoms allow iron to be used in bridges like the one above at Coalbrookedale

oxygen atom

$A_r(RAM) = 16$

symbol O

oxygen molecule

$M_r(RMM) = 32$

formula O_2

Figure 3
An atom and a molecule of oxygen

Oxygen and other non-metals

The relative atomic mass of oxygen is 16, but experiments show that particles of oxygen in the air have a relative mass of 32 (figure 3). This shows that oxygen consists of molecules containing two atoms joined together. The formula of oxygen gas is therefore O_2, not O. We say that O_2 has a **relative molecular mass** (RMM) or **relative formula mass** of 32.

Element	Formula
Oxygen	O_2
Hydrogen	H_2
Nitrogen	N_2
Chlorine	Cl_2
Sulphur	S_8
Argon	Ar

Table 2: The formulas of some non-metals

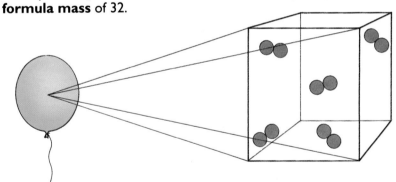

Figure 4
In gaseous oxygen, the molecules of O_2 are widely spaced

*Most non-metals are like oxygen—they have a small number of atoms joined tightly together to form a molecule (table 2). All the common gaseous elements have molecules containing two atoms. They are often described as **diatomic molecules**. There are strong forces between the two atoms in one molecule of oxygen, but very weak forces between one molecule and another. Therefore, the O_2 molecules are easy to separate and oxygen has a very low melting point and boiling point. The molecules are widely-spaced at room temperature (Figure 4) so the density is very low. Since the particles are widely spaced, it is possible to squeeze them into a much smaller volume. This explains why gases can be compressed easily in bagpipes, balloons, tyres and inflatable toys.*

Questions

1 The formula for nitrogen is N_2. What does this mean?
2 How many atoms are there in one molecule of sulphur?
3 Why do metals have a high density?
4 Why does sulphur melt at a much lower temperature than copper?
5 Suppose your best friend has missed the last few lessons. Write down how you would explain to him or her what we mean by (i) an atom, (ii) the difference between an atom and a molecule, (iii) symbols, (iv) the difference between symbols and formulas, (v) relative atomic mass, (vi) relative molecular mass.

9 Particles in Reactions—Equations

The original balance with which Dalton studied the amounts of substances which react together. Dalton suggested that atoms were simply rearranged in chemical reactions

Writing equations

So far, we have used word equations to show what happens when substances react. When magnesium reacts with oxygen, the product is magnesium oxide. We can summarize this in a word equation as:

$$\text{magnesium} + \text{oxygen} \rightarrow \text{magnesium oxide}$$

It would be easier to use formulas instead of names when we write equations. We could write magnesium as Mg, oxygen as O_2, and magnesium oxide as MgO. This gives

$$Mg + O_2 \rightarrow MgO$$

But, notice that this does not balance. There are two oxygen atoms in the O_2 molecule on the left and only one oxygen atom in MgO on the right. So, MgO on the right must be doubled to give

$$Mg + O_2 \rightarrow 2MgO$$

Unfortunately, the equation still does not balance. There are two Mg atoms on the right in 2MgO, but only one on the left in Mg. This can easily be corrected by writing 2Mg on the left, i.e.

$$2Mg + O_2 \rightarrow 2MgO$$

This example shows the three stages in writing an equation:

1 *Write a word equation for the reaction,*

$$\text{e.g. hydrogen} + \text{oxygen} \rightarrow \text{water}$$

2 *Write formulas for the reactants and products,*

$$\text{e.g. } H_2 + O_2 \rightarrow H_2O$$

3 *Balance the equation by making the number of atoms of each element the same on both sides.*

$$\text{e.g. } 2H_2 + O_2 \rightarrow 2H_2O$$

An equation is a summary of the starting substances and the products in a chemical reaction. Equations with formulas help us to see how the atoms are rearranged in a chemical reaction. You can see this even better using models to represent the equation, as in figure 1.

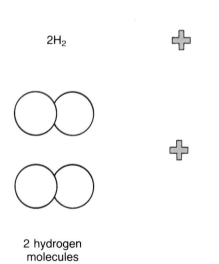
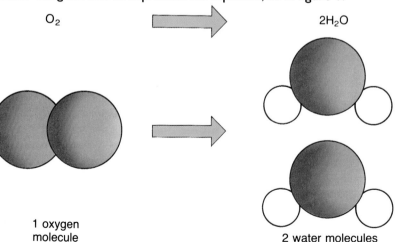

$2H_2$ + O_2 → $2H_2O$

2 hydrogen molecules

1 oxygen molecule

2 water molecules

Figure 1
Using models to represent the reaction between hydrogen and oxygen to form water

The elements hydrogen, oxygen, nitrogen, chlorine, bromine and iodine exist as diatomic molecules. Thus, they are written in equations as H_2, O_2, N_2, Cl_2, Br_2 and I_2. Other elements are shown as single atoms and represented by a monatomic symbol in equations; i.e. Mg for magnesium, Cu for copper, S for sulphur, etc. There is one other important point to remember in writing equations. *Never change a formula* to make an equation balance. For example, the formula of magnesium oxide is always MgO. Mg_2O and MgO_2 do not exist. Equations can only be balanced by putting a number *in front* of the whole formula, i.e. $2MgO$ or $3MgO$.

When natural gas burns, methane (CH_4) reacts with oxygen in the air to form carbon dioxide and water. Write a word equation and then a balanced formula equation for this reaction

The copper dome on Brighton Pavilion is green because the copper has reacted with oxygen in the air to form copper oxide and then this has reacted very slowly with water to form green copper hydroxide. Write word equations and then balanced formula equations for these two reactions

Extra information in equations

● **State symbols** show the state of a substance. The symbol(s) after a formula indicates the substance is a solid; (l) is used for liquid; (g) for gas and (aq) for an aqueous solution (i.e. a substance dissolved in water). For example,

zinc + sulphuric acid → zinc sulphate + hydrogen
$$Zn(s) + H_2SO_4(aq) \rightarrow ZnSO_4(aq) + H_2(g)$$

● **Reaction conditions.** The conditions needed for a reaction can be shown above the arrow in the equation. For example,

hydrogen peroxide $\xrightarrow{\text{MnO}_2 \text{ catalyst}}$ water + oxygen

$$2H_2O_2 \xrightarrow{\text{MnO}_2 \text{ catalyst}} 2H_2O + O_2$$

calcium carbonate $\xrightarrow{\text{heat}}$ calcium oxide + carbon dioxide
(limestone)   (lime)

$$CaCO_3 \xrightarrow{\text{heat}} CaO + CO_2$$

Questions

1 What is an equation?
2 What are the important rules to follow in writing an equation?
3 Look at the photos on this page and answer the questions in the captions.
4 Write word equations and then balanced formula equations for the following reactions;
   (i) aluminium with oxygen to give aluminium oxide (Al_2O_3);
   (ii) copper oxide with sulphuric acid to give copper sulphate ($CuSO_4$) and water;
   (iii) nitrogen with hydrogen to give ammonia (NH_3);
   (iv) charcoal (carbon) burning in oxygen to give carbon dioxide;
   (v) iron with chlorine to give iron chloride ($FeCl_3$).

10 Formulas and Equations

The yellow substance in yellow 'no parking' lines is lead chromate, $PbCrO_4$. What does this formula tell us about lead chromate?

Using formulas

Formulas are useful in several ways.

1 They tell us which elements are present in a compound and the relative numbers of atoms of the different elements. So, the formula CH_4 for methane tells us that methane contains 4 hydrogen atoms for every one carbon atom.

2 They enable us to compare the relative masses of different molecules. For example, using relative atomic masses for hydrogen, carbon and oxygen of 1, 12 and 16 respectively:
the relative molecular mass (relative formula mass) of water,
$H_2O = 1 + 1 + 16 = 18$
the relative formula mass of carbon dioxide,
$CO_2 = 12 + 16 + 16 = 44$
So, carbon dioxide molecules are about $2\frac{1}{2}$ times heavier than water molecules.

3 They enable us to calculate the mass of 1 mole of different substances. For example:
the mass of 1 mole of water, $H_2O = 18\,g$
the mass of 1 mole of calcium carbonate, $CaCO_3$,
$= 40.0 + 12.0 + 16.0 + 16.0 + 16.0 = 100\,g$.

4 They enable us to calculate the masses of elements which combine to form a compound. For example, sulphuric acid, H_2SO_4 is manufactured from sulphur, S.
Thus 1 mole of S (S = 32) gives 1 mole of H_2SO_4.
1 mole of S is 32 g, and
1 mole of H_2SO_4 is $(2 \times 1) + 32 + (4 \times 16) = 98\,g$.
So 32 g of sulphur produce 98 g of sulphuric acid.

This sulphur is being stored after mining. 32 g of sulphur can be reacted with oxygen and water to manufacture 98 g of sulphuric acid

This photo shows one mole of common salt (sodium chloride NaCl), which has a mass of $23 + 35.5 = 58.5\,g$

Using equations

Look at the following equation for the reaction of sodium with chlorine:

$$2Na + Cl_2 \rightarrow 2NaCl$$

This equation tells us that:

1 sodium (Na) reacts with chlorine (Cl_2) to form sodium chloride (NaCl);
2 2 atoms of Na react with 1 molecule of Cl_2 to give 2 particles of NaCl;
3 2 moles of Na react with 1 mole of Cl_2 to give 2 moles of NaCl;
4 Using relative atomic masses (Na = 23.0 and Cl = 35.5);
   2×23 g Na react with 2×35.5 g Cl to give 2×58.5 g NaCl.

> This example shows that equations can tell us
> - the reactants and products
> - the formulas of these substances
> - the formulas of particles (or moles) of the reactants and products
> - the masses of the reactants and products.

It is useful in industry to know the amounts of reactants and products. Industrial chemists need to know how much product they can make from a given amount of starting material. For example, using an equation and relative atomic masses, we can calculate how much lime (calcium oxide) can be obtained from 1 kg of pure limestone (calcium carbonate).

calcium carbonate $\rightarrow$ calcium oxide + carbon dioxide
$$CaCO_3 \rightarrow CaO + CO_2$$
$\therefore$ 1 mole $CaCO_3 \rightarrow$ 1 mole CaO
$(40 + 12 + (3 \times 16))$ g $CaCO_3 \rightarrow (40 + 16)$ g CaO
100 g $CaCO_3 \rightarrow$ 56 g CaO
$\therefore$ 1000 g $CaCO_3 \rightarrow$ 560 g CaO

So 1 kg of pure limestone produces 0.56 kg of lime.

Using equations, it is also possible to calculate the volumes of gases which react. Chemists have found that *1 mole of any gas occupies 24 dm^3 at room temperature (20° C) and atmospheric pressure.* (At *standard* temperature and pressure, or s.t.p., (0° C and 1 atmosphere), 1 mole of a gas occupies (22.4 dm^3.)

For example, ammonia (NH_3) is manufactured from nitrogen (N_2) and hydrogen (H_2).

	nitrogen	+	hydrogen	$\rightarrow$	ammonia
	N_2	+	$3H_2$	$\rightarrow$	$2NH_3$
$\therefore$	1 mole N_2	+	3 moles H_2	$\rightarrow$	2 moles NH_3
i.e.	24 dm^3 nitrogen	+	3×24 dm^3 hydrogen	$\rightarrow$	2×24 dm^3 ammonia
So	1 dm^3 nitrogen	+	3 dm^3 hydrogen	$\rightarrow$	2 dm^3 of ammonia

The lime kiln converts limestone into lime

Questions

1 60 g of a metal M (M = 60) combine with 24 g of oxygen (O = 16).
   (a) How many moles of O react with one mole of M?
   (b) What is the formula of the oxide of M?

2 The fertilizer 'nitram' (ammonium nitrate) has the formula NH_4NO_3.
   (a) What are the masses of nitrogen, hydrogen and oxygen in 1 mole of nitram?
   (N = 14, H = 1, O = 16)
   (b) What are the percentages of nitrogen, hydrogen and oxygen in nitram?

3 Nitram is made from ammonia and nitric acid.

   NH_3 + HNO_3 $\rightarrow$ NH_4NO_3
   ammonia nitric acid ammonium nitrate

   (a) What does this equation tell you?
   (b) What mass of NH_3 reacts with HNO_3 to give 1 mole of Nitram? (N = 14, H = 1, O = 16)
   (c) What mass of NH_3 reacts with HNO_3 to give 1 kg of Nitram?
   (d) What volume of NH_3 at s.t.p. produces 1 mole of Nitram?

Section C: Activities

1 Soluble aspirin

When a soluble aspirin tablet is added to water, bubbles start to form. The mass of the solution after the tablet has dissolved is less than the mass of the tablet plus the mass of the water before mixing.

1 Why do you think there is a loss in mass when the soluble aspirin is added to water?

2 Describe an experiment which you could carry out to test your ideas for the loss in mass. (Draw diagrams if it will help.) You should try to use materials and apparatus which are readily available in the lab.

3 What do you think happens to particles in the soluble aspirin when it is added to water?

4 Soluble aspirins dissolve faster in warm water than in cold water. Why is this?

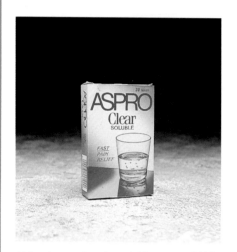

Why do people take soluble aspirins? Why do the tablets fizz when added to water?

2 Extracting tin from tinstone

The Production Manager at a tin smelter has to meet certain production targets. The smelter must produce 595 tonnes of tin each month. In order to do this, the manager must calculate how much purified tinstone (tin oxide) he must buy to produce 595 tonnes of tin. The basis for his calculation is the equation

tin oxide	+	coke (carbon)	→	tin	+	carbon monoxide
SnO_2	+	$2C$	→	Sn	+	$2CO$

SnO_2 means that tin oxide contains 2 atoms of oxygen for every 1 atom of tin.

CO means that carbon monoxide contains 1 atom of carbon for every 1 atom of oxygen.

Using relative atomic masses, the manager can now calculate how much tin can be obtained from a certain amount of tin oxide (SnO_2). The relative formula masses of SnO_2 and CO can be calculated by adding together the correct number of relative atomic masses for the atoms of the elements in each compound.

Relative atomic mass of tin = 119
Relative atomic mass of oxygen = 16
Relative atomic mass of carbon = 12
∴ the relative formula mass of SnO_2 = 119 + (2 × 16) = 151

1 What is the relative formula mass of CO?

When we put relative masses below the formulas in the equation, we get:

$$SnO_2 + 2C → Sn + 2CO$$
Relative masses: 151 + ? → 119 + ?

This means that 151 tonnes of tin oxide (tinstone) can produce 119 tonnes of tin.

If the Production Manager had a monthly production target of 119 tonnes of tin, then 151 tonnes of tinstone are required.

2 What mass of tinstone is needed to meet the production target of 595 tonnes of tin per month?

3 What mass of coke (carbon) is needed each month?

4 The smelter normally operates all night and day. It produces about 1 tonne of tin per hour. Is the production target possible? Explain your answer.

5 In 1993, tin from the smelter was sold at £5000 per tonne. Suppose the smelter produces 600 tonnes of tin in December. What is the value of this tin?

6 What concerns will the Production Manager have regarding the tin smelter and environmental issues?

3   Finding the formula of red copper oxide

Jason and Meera had just completed their experiment with black copper oxide. Their results showed that its formula was CuO. After the experiment, Jason read about a second oxide of copper called red copper oxide. So he and Meera decided to investigate its formula.

    They took a weighed amount of red copper oxide and reduced it to copper. In order to get a more reliable result, they carried out the experiment five times, starting with different amounts of red copper oxide. Their results are shown in Table 1.

Recording the results

1 Start a spreadsheet program on a computer and open up a new spreadsheet for your results.

2 Enter the experiment numbers and the masses of copper oxide and copper as in Table 1.

3 Enter a formula in column 4 to work out the mass of oxygen in the red copper oxide taken. (If you cannot do this, ask your teacher to help you type it in.)

4 Enter a formula in column 5 to find the number of moles of copper in the oxide (Cu = 63.5).

5 Enter a formula in column 6 to find the number of moles of oxygen in the oxide (O = 16).

6 From the spreadsheet, plot a line graph of moles of copper (y-axis) against moles of oxygen (x-axis). (If you cannot pick graphs directly from the spreadsheet, draw the graphs by hand.)

7 Look at your graph.

(a) What is the average value for $\dfrac{\text{moles of copper}}{\text{moles of oxygen}}$ in red copper oxide?

(b) How many moles of copper combine with one mole of copper in red copper oxide?

(c) What is the formula for red copper oxide?

8 If possible, print a copy of your spreadsheet and your graph.

The disused tin mine at Wheal Coates, near St Agnes in Cornwall. Why did the mines have tall chimneys? Why is tinstone no longer mined in Cornwall?

Expt No.	Mass of red copper oxide taken/g	Mass of copper in the oxide/g
1	1.43	1.27
2	2.10	1.87
3	2.86	2.54
4	3.55	3.15
5	4.29	3.81

Table 1: The results of experiments to investigate the formula of red copper oxide

Section C: Study Questions

1 The kinetic theory of matter states that all substances contain particles which are moving. Use this theory to explain the following:
  (a) When a balloon is taken out of doors on a cold day, it gets smaller.
  (b) Water boils at less than 80° C on the top of Mount Everest.
  (c) Snow and ice sometimes disappear on a cold winter's day without leaving any trace of water.
  (d) Solids, such as salt and sugar, can dissolve in water.
  (e) Dogs can be trained to find hidden drugs such as cannabis.

2 (a) What name is used for the smallest particle of a compound which can freely exist?
  (b) What name is used for the smallest particle of an element which has the properties of that element?
  (c) Which particles in an atom are used in chemical bonding?
  (d) Sometimes we talk about 'splitting the atom' to create a new element. What part of the atom is actually 'split'?
  (e) If the formula for water is H_2O, how many atoms are there in one molecule of water?
  (f) Write the formula for six molecules of water.

3 In the boxes below, ⚪ and ⚪ represent different atoms.

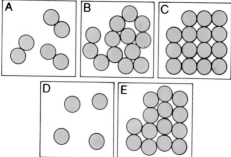

Which box contains
  (a) a close packed metal,
  (b) a solid compound,
  (c) a liquid,
  (d) a mixture,
  (e) a diatomic molecule?

4 What atoms and molecules do the following contain? Air, acetic acid (CH_3COOH) in vinegar, sulphuric acid (H_2SO_4), limestone ($CaCO_3$).

5 Copper carbonate decomposes, when it is heated, into copper oxide and carbon dioxide according to the equation

$$CuCO_3 \rightarrow CuO + CO_2$$

An empty test tube was first weighed. Some copper carbonate was put in it, and it was weighed again. The diagrams below show the scale of the top-pan balance for these weighings.

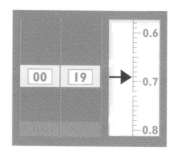
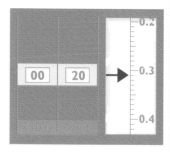

(a) Complete the results table below.

Mass of empty test tube	___ g
Mass of test tube + copper carbonate	___ g
Mass of copper carbonate taken	___ g

(b) Calculate the mass of 1 mole of
(i) copper carbonate,
(ii) copper oxide (Cu = 64, C = 12, O = 16).
(c) Calculate the expected mass of copper oxide remaining in the experiment in which the test tube was heated.
(d) The test tube was allowed to cool before it was reweighed. The mass remaining was found to be 0.45 g. Compare this result with the expected result obtained in part (c) and account for any difference.
(e) Why was the test tube allowed to cool before it was weighed?
(f) What would you do to ensure that all the copper carbonate had decomposed?

6 A solid compound contains 39% potassium, 1% hydrogen, 12% carbon and 48% oxygen, by weight.
  (a) What is the ratio of the weights of K:H:C:O in the compound?
  (b) What is the ratio of moles of K:H:C:O in the compound?
  (c) What is the formula of the solid?

7 Tinstone, SnO_2, is an important mineral from which tin can be obtained.
  (a) How much does 1 mole of tinstone weigh?
  (b) How much tin does 1 mole of tinstone contain?
  (c) What is the percentage of tin in tinstone?
  (d) How much tin is there in 5000 kg of tinstone?
  (e) Where are the world's major deposits of tinstone found? (You will need to refer to an encyclopaedia.)
  (f) When tinstone is converted into tin, is the tinstone *oxidized*, *reduced*, *synthesized*, or *decomposed*?

8 (a) Rust is hydrated iron oxide. Its formula can be written as $Fe_xO_y \cdot ZH_2O$. Describe an experiment you could do to find the values of x, y and z.
  (b) A sample of rust contained 56.0 g of iron, 24.0 g of oxygen and 27.0 g of water. What is the formula of this rust?

Electricity and Electrolysis

Electricity is transmitted across the country from power stations to our homes, schools and factories. Sometimes the pylons and cables spoil the countryside. Why are the cables not buried below ground?

1 Electricity in Everyday Life

A city at night. How could we survive without electricity?

Look at all the electrical appliances in this kitchen. It is difficult to imagine life without electrical appliances

It is hard to imagine life without electricity. Every day we depend on electricity for cooking, for lighting and for heating. At the flick of a switch, we use electric fires, electric kettles and dozens of other electrical gadgets. All these electrical appliances use mains electricity. The electricity is generated in power stations from coal, oil or nuclear fuel. Heat from the fuel is used to boil water. The steam produced drives turbines and generates electricity. In this way, chemical energy in the fuel is converted into electrical energy.

In addition to the many appliances which use mains electricity, there are others like torches, radios and calculators that use electrical energy from cells and batteries. Electricity can also be used to manufacture some important chemicals. For example, salt (sodium chloride) cannot be decomposed into sodium and chlorine using heat, but it can be decomposed using electricity. Sodium and chlorine are manufactured by passing electricity through molten sodium chloride mixed with a little calcium chloride at 600° C.

$$\text{sodium chloride} \xrightarrow{\text{electricity}} \text{sodium} + \text{chlorine}$$

These uses of electricity show why it is so useful to our society.
- *It can be used to transfer energy easily from one place to another.*
- *It can be converted into other forms of energy and used to warm a room, light a torch or cook a meal.*

Electric charges

Comb your hair quickly with a plastic comb and then use the comb to pick up tiny bits of paper. Why does this happen?

Atoms are composed of three particles—**protons**, **neutrons** and **electrons**. The centre of the atom, called the **nucleus**, contains protons and neutrons. Protons are positive but neutrons have no charge. Electrons occupy the outer parts of the atom and move around the nucleus. The negative charge on one electron just balances the positive charge on one proton so atoms have equal numbers of protons and electrons. For example, hydrogen atoms have 1 proton and 1 electron. Carbon atoms have 6 protons, 6 electrons and 6 neutrons. Aluminium atoms have 13 protons, 13 electrons and 14 neutrons (Figure 1).

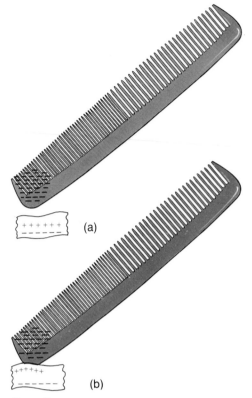

(a)

(b)

Figure 2

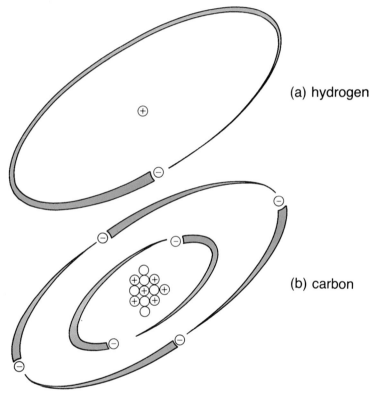

(a) hydrogen

(b) carbon

Figure 1
Protons, neutrons and electrons in a hydrogen atom and a carbon atom
($\oplus$ = proton, $\bigcirc$ = neutron, $\ominus$ = electron)

When you comb your hair quickly with a plastic comb, the comb pulls electrons off the atoms in your hair. The comb will then have more electrons than protons so it has a negative charge overall. Your hair will have fewer electrons than protons and so it has a positive charge overall.

When you bring the charged comb close to tiny pieces of paper, the negative charge on the comb repels electrons from the area of the paper nearest to it (figure 2(a)). This part of the paper therefore becomes positive and it is attracted to the comb because *unlike charges attract*. If the paper is light enough, it can be picked up (figure 2(b)).

The plastic comb has just been rubbed on a woollen jumper. What is making the water bend?

Questions

1 Electricity is used more widely than gas for our energy supplies. Why is this?

2 (a) List 4 important uses which electricity has in your home.
 (b) How would the jobs in part (a) be done without electricity?

3 Lithium atoms have 3 protons, 3 electrons and 4 neutrons. Draw a picture of a lithium atom similar to that for a carbon atom in figure 1.

4 Look at the photo above. Why does the water bend?

2 Electric Currents

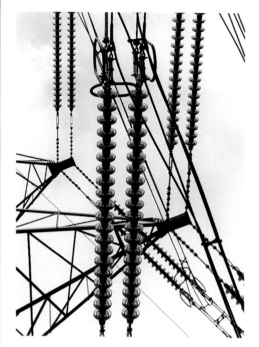

Plastic and glass insulators are used to hold the cables to pylons in the Grid System

Materials, such as plastic combs, which hold their charge on rubbing and do not allow electrons to pass through them are called **insulators**. Plastics like polythene, cellulose acetate and PVC are used to insulate electrical wires and cables. Materials like metals, that allow electrons to pass through them are called **conductors** (see the table).

Conductors			Insulators
Good	**Moderate**	**Poor**	

metals, e.g.
    copper
    aluminium
    iron
	carbon (graphite)
	silicon
	germanium
		water	rubber
			air
			plastics, e.g
			    polythene
			    PVC

Some conductors and insulators

Thus, copper, iron and aluminium, which are easily made into wire, are used for fuses, wires and cables in electrical machinery. Copper is used more than any other metal in electrical wires and cables because electrons move through it easily—it is a good conductor. When an electric current flows through the wire, electrons in the copper atoms are attracted towards the positive terminal of the battery. At the same time, extra electrons are repelled into the copper wire from the negative terminal (figure 1). *An electric current is simply a flow of electrons.* Electrons flow through the metal rather like water flows through a pipe or traffic moves along a road.

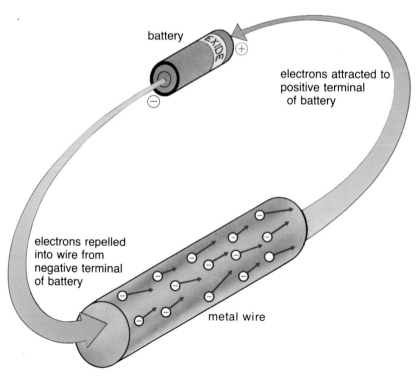

The cut-away end of this electrical cable shows the layers of different materials which make it up. Bundles of copper wire are at the centre. Copper is used because it is such a good conductor. The layers of material around the copper provide insulation and protection

battery

electrons attracted to positive terminal of battery

electrons repelled into wire from negative terminal of battery

metal wire

Figure 1

In addition to insulators and conductors, there are a few substances called *semiconductors*. Silicon and germanium containing a trace of impurity are two of the best known semiconductors. These substances allow electrons to flow easily through them in one direction, but not in the other. Semiconductors of this kind form an important part of any transistor. In some cases, hundreds of these tiny transistors are built up on a small flat plate forming a silicon chip.

A silicon chip

Measuring electricity

We could measure the amount of electricity (electric charge) that has flowed along a wire by counting the number of electrons which pass a certain point. The charge on one electron is, however, much too small to be used as a practical unit in measuring the quantity of electricity. The practical unit normally used is the **coulomb** (C), which is about six million, million, million (6×10^{18}) electrons.

If one coulomb of charge passes along a wire in one second, then the rate of charge flow (i.e. the electric current) is 1 coulomb per second or 1 **ampere** (A). If 3 coulombs pass along the wire in 2 seconds, then the current is $\frac{3}{2}$ coulombs per sec or $\frac{3}{2}$ A. If Q coulombs flow along a wire in t seconds, the electric current (I) is given by,

$$I = \frac{Q}{t}$$

This equation can be rearranged to give

$$Q = I \times t$$

i.e. $\quad \begin{matrix} \text{charge} \\ \text{in coulombs} \end{matrix} = \begin{matrix} \text{current} \\ \text{in amps} \end{matrix} \times \begin{matrix} \text{time} \\ \text{in seconds} \end{matrix}$

$\therefore$ 1 A for 1 sec = $1 \times 1 = 1$ C
2 A for 1 sec = $2 \times 1 = 2$ C
2 A for 2 sec = $2 \times 2 = 4$ C

The equation charge = current $\times$ time can be compared to the flow of water along a pipe since:

amount of water passed = rate of flow of water $\times$ time
(i.e. current)

Questions

1 What is (i) an electric current; (ii) a conductor; (iii) an insulator?

2 (a) Name 4 elements which conduct electricity when solid.
(b) Name 4 elements which conduct electricity when liquid.
(c) Name 4 elements which do not conduct electricity when solid.

3 Design an experiment to compare the conduction of electricity by thin copper wire and thick copper wire.
(a) Draw a diagram of the apparatus you would use.
(b) Say what you would do.
(c) Say what measurements you would make.
(d) Say how you would compare the two wires.

4 The current in a small torch bulb is 0.25 A. How much electricity flows if the torch is used for 15 minutes?

5 The current in a car headlamp is 4 A.
(a) What is the current in coulombs per sec?
(b) What is the current in electrons per sec? (Assume 1 coulomb = 6×10^{18} electrons.)
(c) How much electricity (electric charge) flows if the headlamp is used for 1 hour?

6 Look at the photo of the silicon chip above. Find out more about silicon chips and the transistors and semiconductors in them.

3 Which Substances Conduct Electricity?

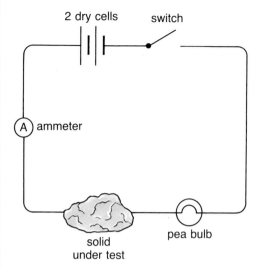

Figure 1

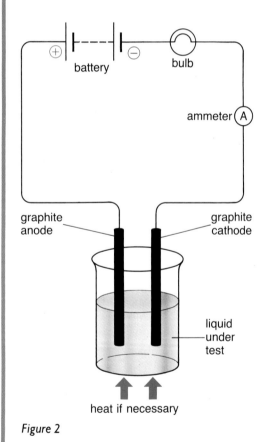

Figure 2

Which solids conduct?

The apparatus in figure 1 can be used to test whether a solid conducts electricity. If the solid conducts, what happens when the switch is closed?

Experiments show that *the only common solids which conduct electricity well are metals and graphite.* When metals and graphite conduct electricity, electrons flow through the material, but there is no chemical reaction. No new substances form and there is no change in weight. No solid *compounds* conduct electricity.

Which liquids conduct?

Pure water does not conduct electricity. But, water containing a little sulphuric acid will conduct electricity. Unlike metals, the water changes when it conducts. It is decomposed into hydrogen and oxygen. Electricity is a form of energy, like heat. We can use it to boil water, to cook food and to cause chemical reactions.

The decomposition of a substance, such as water, by electricity is called **electrolysis**. The compound which is decomposed is called an **electrolyte** and we say that it has been **electrolysed**.

Figure 2 shows how we can test the conductivity of liquids. The terminals through which the current enters and leaves the electrolyte are called **electrodes**. The electrode connected to the positive terminal of the battery is positive itself and is called the **anode**. The electrode connected to the negative terminal of the battery is negative itself and is called the **cathode**. Table 1 shows the results of tests on various liquids and aqueous solutions.

Pure liquids	Does the liquid conduct?	Aqueous solutions	Does the solution conduct?
Bromine	No	Ethanol (C_2H_6O)	No
Mercury	Yes	Sugar ($C_{12}H_{22}O_{11}$)	No
Molten sulphur	No	Sulphuric acid (H_2SO_4)	Yes
Molten zinc	Yes	Acetic acid ($C_2H_4O_2$)	Yes
Water (H_2O)	No	Copper sulphate ($CuSO_4$)	Yes
Ethanol (C_2H_6O)	No	Sodium chloride (NaCl)	Yes
Tetrachloromethane (CCl_4)	No	Potassium iodide (KI)	Yes
Molten sodium chloride (NaCl)	Yes		
Molten lead bromide ($PbBr_2$)	Yes		

Table 1: Testing to see which liquids and aqueous solutions conduct

Aluminium is manufactured by electrolysis from aluminium oxide. The workman is tapping molten aluminium from the cell where electrolysis occurs

Look at the results in table 1.

1 Do the liquid metals conduct electricity?
2 Do the liquid non-metals conduct electricity?
3 Do the compounds containing only non-metals (non-metal compounds) conduct (i) when liquid; (ii) in aqueous solution?
4 Do the compounds containing both metals and non-metals (metal/non-metal compounds) conduct electricity (i) when liquid; (ii) in aqueous solution?

The answers to these questions and the results of the experiment are summarized in table 2.

Substance	Elements		Compounds	
	Metals and graphite	Non-metals except graphite	Metal/non-metal	Non-metal
Examples	Fe, Zn	Br_2, S	NaCl, $CuSO_4$	C_2H_6O, CCl_4
Solid	Yes	No	No	No
Liquid	Yes	No	Yes	No
Aqueous solution	—	—	Yes (and acids)	No (except acids)

Table 2: The conduction of electricity by elements and compounds

Notice the following points from these results:

• *Metal/non-metal compounds conduct electricity when they are molten (liquid) and when they are dissolved in water (aqueous). These compounds are decomposed during electrolysis.*

• *Non-metal compounds do not conduct in the liquid state or in aqueous solution (except aqueous solutions of acids)*

Questions

1 Explain the following words: *electrolysis; electrolyte; electrode; anode; cathode.*
2 *Calcium; carbon disulphide; copper sulphate solution; lead; carbon; water; methane (natural gas); phosphorus; dilute sulphuric acid.*
From this list name: (i) two metals; (ii) two non-metals; (iii) two electrolytes; (iv) two pure liquids at 20° C; (v) two elements which conduct; (vi) two compounds that are gases at 110° C; (vii) three compounds that are non-electrolytes.
3 Michael Faraday was one of the first scientists to investigate electrolysis. Find out about his life and work. Prepare a short talk about Faraday for the rest of your class.

Michael Faraday — one of the first scientists to investigate electrolysis

4 Investigating Electrolysis

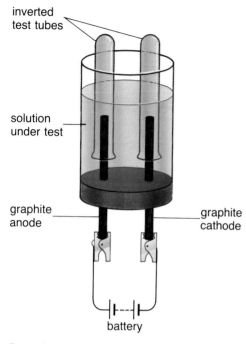

inverted
test tubes

solution
under test

graphite
anode

graphite
cathode

battery

Figure 1

Products of electrolysis

When compounds are electrolysed, new substances are produced at the electrodes. For example, when electricity is passed through molten sodium chloride, pale green chlorine gas comes off at the anode and sodium forms at the cathode. When copper sulphate solution is electrolysed using the apparatus in figure 1, a pink deposit of copper appears on the cathode. Bubbles of a colourless gas stream off the anode and collect in the inverted test tube. This gas relights a glowing splint. This shows that it is oxygen.

The table lists the products formed at the electrodes when various liquids and aqueous solutions are electrolysed. Remember that water in the aqueous solutions may be electrolysed.

Substance electrolysed	Product at anode	Product at cathode
Molten lead bromide	Brown fumes of bromine	Lead
Molten sodium chloride	Pale green chlorine gas	Sodium
Aqueous potassium iodide	Iodine which colours the solution brown	Hydrogen
Aqueous copper sulphate	Oxygen	Copper (deposited on the cathode)
Hydrochloric acid	Chlorine	Hydrogen
Aqueous zinc bromide	Bromine which colours the solution brown	Zinc (deposited on the cathode)
Aqueous copper chloride	Chlorine	Copper (deposited on the cathode)

The products formed at the electrodes when some liquids and aqueous solutions are electrolysed.

1 Which elements are produced at the anode?
2 Which elements are produced at the cathode?

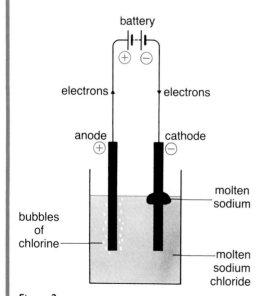

battery

electrons

electrons

anode

cathode

bubbles
of
chlorine

molten
sodium

molten
sodium
chloride

Figure 2

When acids and metal/non-metal compounds conduct electricity
- *metals or hydrogen are formed at the cathode*
- *non-metals (except hydrogen) are formed at the anode*

These compounds are decomposed by electrical energy and an element is produced at each electrode. This is very different from the conduction of electricity by metals which are not decomposed during conduction. The first two electrolyses in the table (left) can be summarized in word equations as:

$$\text{lead bromide} \xrightarrow{\text{electrical energy}} \text{lead} \quad + \text{bromine}$$

$$\text{sodium chloride} \xrightarrow{\text{electrical energy}} \text{sodium} \quad + \text{chlorine}$$

Explaining electrolysis

Sodium and chlorine are manufactured by electrolysis of molten sodium chloride. When an electric current passes through molten sodium chloride, a shiny bead of sodium is produced at the cathode and chlorine gas forms at the anode (figure 2). This decomposition is caused by electrical energy in the current, but how does this happen? Sodium particles in the electrolyte must be positive as they are attracted to the negative cathode. At the same time, chlorine is produced at the anode, so chloride particles in the electrolyte are probably negative.

The formula of sodium chloride is NaCl so we can think of this as positive Na^+ particles and negative Cl^- particles. Since NaCl is neutral, the positive charge on one Na^+ must balance the negative charge on one Cl^-. These charged particles, like Na^+ and Cl^-, which move to the electrodes during electrolysis are called **ions**.

During electrolysis, Na^+ ions near the cathode combine with negative electrons on the cathode forming neutral sodium atoms (figure 3a):

Na^+	+	e^-	⟶	Na
sodium ion in sodium chloride electrolyte		electron on cathode from battery		sodium atom in metal

At the anode, Cl^- ions lose an electron to the positive anode leaving neutral chlorine atoms (figure 3b):

Cl^-	⟶	e^-	+	Cl
chloride ion in electrolyte		electron given to anode		chlorine atom

The Cl atoms then join up in pairs to form molecules of chlorine gas, Cl_2:

Cl	+	Cl	⟶	Cl_2
chlorine atom		chlorine atom		chlorine molecule

These equations show that Na^+ ions remove electrons from the cathode, and Cl^- ions give up electrons to the anode during electrolysis. The electric current is being carried through the molten sodium chloride by ions. The electrolysis of other molten and aqueous substances can also be explained in terms of ions.

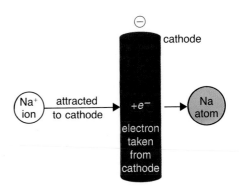

Figure 3(a)

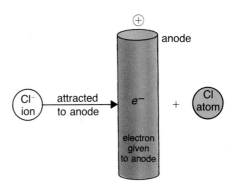

Figure 3(b)

Questions

Write *true* or *false* to each answer in questions 1 to 3.
1 When pure lead bromide is electrolysed
A it must be molten
B lead forms at the anode
C decomposition occurs
D a brown gas forms at the anode
E the process is exothermic.
2 The following substances are electrolytes:
A copper sulphate solution
B sugar solution
C dilute sulphuric acid
D copper
E molten wax
F liquid sulphur.
3 When potassium chloride solution is electrolysed, the products include:
A chlorine at the anode
B hydrogen at the cathode
C oxygen at the cathode
D potassium at the cathode
E chlorine at the cathode
F potassium at the anode.
4 Plan an experiment to coat a steel paper clip with copper.

5 Charges on Ions

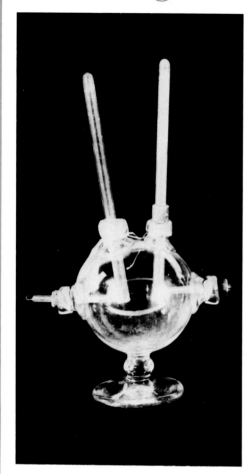

The electrolysis equipment in which Faraday studied the charges on ions. Where are the electrodes? What are the vertical tubes for?

During electrolysis
- *a metal or hydrogen forms at the negative cathode*
- *a non-metal (except hydrogen) forms at the positive anode*

The negative cathode attracts positive ions and the positive anode attracts negative charges. So we can deduce that:

- *Metals and hydrogen have positive ions. These ions are called* **cations** *because they are attracted to the cathode.*
- *Non-metals (except hydrogen) have negative ions. These ions are called* **anions** *because they are attracted to the anode.*

Electrolysis experiments show that it requires twice as much electricity (twice as many electrons) to produce a magnesium atom as it does to produce a sodium atom. The formation of sodium during electrolysis can be written as

$$Na^+ \quad + \quad e^- \quad \rightarrow \quad Na$$
sodium ion   one electron   sodium atom

So the formation of magnesium during electrolysis is written as

$$Mg^{2+} \quad + \quad 2e^- \quad \rightarrow \quad Mg$$
magnesium ion   two electrons   magnesium atom

The experiments and equations show that a sodium ion can be written as Na^+ and a magnesium ion as Mg^{2+}.

In this way we can build up a list of ions with their charges, like those in the table (right). Notice that copper can form two ions, Cu^+ and Cu^{2+}. We show this in the names of its compounds by using the names copper(I) and copper(II). Thus copper forms two oxides, two chlorides, two sulphates, etc. The correct names for its two oxides are copper(I) oxide which is red, and copper(II) oxide which is black. Most of the common copper compounds are copper(II) compounds. These include copper(II) oxide and blue copper(II) sulphate. Iron can also form two different ions, Fe^{2+} and Fe^{3+}, and we use the names iron(II) and iron(III) for their respective compounds.

Most metal ions have a charge of 2+. All the *common* metal ions without a charge of 2+ are shown in the table. These are Ag^+, Na^+ and K^+ with a charge of 1+ (to remember this say 'AgNaK') and Cr^{3+}, Al^{3+} and Fe^{3+} with a charge of 3+ (to remember this say 'CrAlFe').

Copper has two oxides and two chlorides

copper (I) oxide   copper (II) oxide

copper (I) chloride   copper (II) chloride

Farmers often put lime on the soil of their fields. Lime is an ionic compound. Why is it added to the soil?

Notice also in the table below that some negative ions are made from a group of atoms. For example, nitrate, NO_3^-, contains one nitrogen atom and three oxygen atoms. Groups of atoms like this are called **radicals**.

Positive ions (cations)		Negative ions (anions)	
Hydrogen	H^+	Chloride	Cl^-
Sodium	Na^+	Bromide	Br^-
Potassium	K^+	Iodide	I^-
Silver	Ag^+	Nitrate	NO_3^-
Copper(I)	Cu^+	Hydroxide	OH^-
Copper(II)	Cu^{2+}	Oxide	O^{2-}
Magnesium	Mg^{2+}	Carbonate	CO_3^{2-}
Calcium	Ca^{2+}	Sulphide	S^{2-}
Zinc	Zn^{2+}	Sulphate	SO_4^{2-}
Iron(II)	Fe^{2+}	Sulphite	SO_3^{2-}
Iron(III)	Fe^{3+}		
Aluminium	Al^{3+}		
Chromium	Cr^{3+}		

Common ions and their charges

Bonds between ions

Metal/non-metal compounds which are made of ions are called **ionic compounds**. They include common salt (sodium chloride), lime (calcium oxide) and iron ore (iron(III) oxide).

In ionic compounds, the charges on the positive ions just balance the charges on the negative ions. So, because the sodium ion is Na^+ and the chloride ion is Cl^-, we can predict that the formula of sodium chloride will be Na^+Cl^- or simply NaCl. By balancing the charges in this way, we can work out the formulas of other ionic compounds. For example, the formula of lime is $Ca^{2+}O^{2-}$ (CaO) and that of iron(III) oxide is $(Fe^{3+})_2(O^{2-})_3$ or simply Fe_2O_3. In iron(III) oxide, the six positive charges on two Fe^{3+} ions have been balanced by six negative charges on three O^{2-} ions.

Questions

1 Suppose your best friend has been absent from school. Write down what you would say to him or her to explain why an aluminium ion is written as Al^{3+}, whereas a silver ion is written as Ag^+.

2 Look at Faraday's electrolysis equipment in the photograph on the opposite page.
   (a) Where are the electrodes?
   (b) What are the vertical tubes for?
   (c) What do you think would happen if copper(II) chloride solution was electrolysed in the equipment?
   (d) How has our equipment for electrolysis changed since Faraday's time?

3 Write the symbols for the ions in the following compounds and then work out their formulas:
calcium chloride; copper(II) sulphide; aluminium oxide; potassium hydroxide; iron(III) iodide; chromium bromide.

6 Electrolysis in Industry

Obtaining metals by electrolysis

Reactive metals like sodium, magnesium and aluminium cannot be obtained by reducing their oxides to the metal with carbon (coke). These metals can only be obtained by electrolysis of their molten (fused) compounds. We cannot use electrolysis of their aqueous compounds because hydrogen (from the water), and *not* the metal, is produced at the cathode. For example, during the electrolysis of aqueous sodium chloride, hydrogen (not sodium) is produced at the cathode.

Metals low in the reactivity series, such as copper and silver, can be obtained by reduction of their compounds or by electrolysis of their aqueous compounds. When their aqueous compounds are electrolysed, the metal is produced at the cathode rather than hydrogen (from the water).

Manufacturing aluminium by electrolysis

Aluminium is manufactured by the electrolysis of *molten* aluminium oxide. This is obtained from bauxite.

Pure aluminium oxide cannot be used as the electrolyte because it does not melt until 2045° C. This makes its electrolysis uneconomic. The aluminium oxide is therefore dissolved in molten cryolite (Na_3AlF_6) which melts below 1000° C. Figure 1 shows a diagram of the electrolytic method used. Aluminium ions in the electrolyte are attracted to the carbon cathode lining the tank. Here they accept electrons to form aluminium:

Cathode (−) $\qquad\qquad Al^{3+} + 3e^- \rightarrow Al$

Aluminium is obtained from bauxite, impure aluminium oxide. The picture shows bauxite being mined

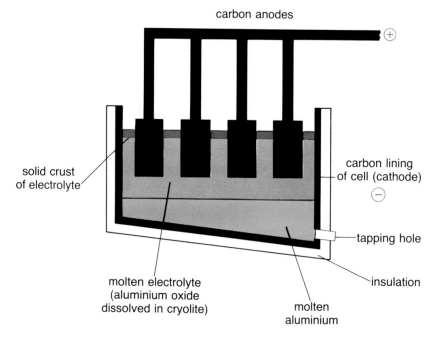

Figure 1
The electrolytic cell for aluminium manufacture

Molten aluminium collects at the bottom of the cell and is tapped off at intervals. It takes about 16 kilowatt-hours of electricity to produce 1 kg of aluminium. The extraction plants are therefore sited near sources of cheap electricity.

Oxide ions (O^{2-}) are attracted to the carbon anodes. Here they give up their electrons forming oxygen atoms:

Anode (+) $\qquad\qquad O^{2-} \rightarrow O + 2e^-$

The oxygen atoms then combine in pairs to form oxygen gas (O_2).

Purifying copper by electrolysis

When copper sulphate solution is electrolysed with copper electrodes, copper is deposited on the cathode and the copper anode loses weight (figure 2).

The aqueous copper sulphate contains copper ions (Cu^{2+}) and sulphate ions (SO_4^{2-}). During electrolysis, Cu^{2+} ions are attracted to the cathode where they gain electrons and deposit on the cathode:

Cathode (−) $\qquad\qquad Cu^{2+} + 2e^- \rightarrow Cu$

SO_4^{2-} ions are attracted to the anode, but they are not discharged. Instead, copper atoms, which make up the anode, give up two electrons each and go into solution as Cu^{2+} ions:

Anode (+) $\qquad\qquad Cu \rightarrow Cu^{2+} + 2e^-$

The overall result of this electrolysis is that the anode loses weight and the cathode gains weight—copper metal is transferred from the anode to the cathode.

This method is used industrially to purify crude copper. The impure copper is the anode of the cell. The cathode is a thin sheet of pure copper. The electrolyte is copper sulphate solution. The impure copper anode dissolves away and pure copper deposits on the cathode.

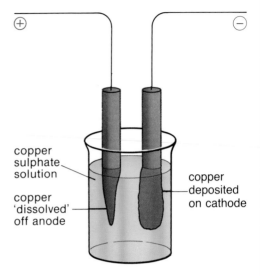

copper sulphate solution

copper 'dissolved' off anode

copper deposited on cathode

Figure 2
Purifying copper by electrolysis

Impure copper anodes being transferred to an electrolysis tank for purification

Questions

1 Get into a group with 2 or 3 others. Suggest reasons for the following.
  (a) Clay is the most abundant source of aluminium, but the metal is never extracted from clay.
  (b) Aluminium extraction plants are usually sited near hydroelectric plants.

2 *The costs of manufacturing aluminium*
It takes about 16 units (kilowatt-hours) of electricity to produce 1 kg of aluminium.
  (a) Suppose aluminium manufacturers pay 3p a unit for electricity. How much is the cost of electricity in producing 1 kg of aluminium?
  (b) What other costs are involved in manufacturing aluminium besides the cost of electricity?

3 Magnesium is manufactured by electrolysis of molten magnesium chloride using a graphite anode and a steel cathode.
  (a) Give the symbols and charges on the ions in the electrolyte.
  (b) Draw a circuit diagram to show the directions in which the ions and electrons move during electrolysis.
  (c) Write equations for the processes which occur at the anode and the cathode.
  (d) Why do you think the anode is made of graphite and *not* steel which is cheaper?

7 Electroplating

This picture shows the bodywork of a car rising out of an electroplating bath. Notice the wire carrying the current from the bodywork

The method for purifying impure copper described in the last unit can be used industrially to coat (plate) articles with copper. The process is called *electroplating*. Several metals can be used for electroplating articles. The most commonly used metals, apart from copper, are chromium, silver and tin. The article to be plated is made the cathode of the cell, the anode is a piece of copper and the electrolyte is copper sulphate solution. During the electroplating process, the copper anode dissolves away and copper deposits on the article at the cathode.

Although it is easy to deposit a metal during electrolysis, the conditions must be carefully controlled so that the metal sticks to the object to be plated. For example, chromium will not stick to iron (steel) or copper during electroplating. Thus, steel articles to be chromium plated (such as bath taps and electric kettles) are first plated with nickel. This forms a firm deposit on the steel which can then be plated with chromium.

These three objects have been electroplated with either copper or chromium

In order to obtain a good coating of metal during electroplating:

1 the object to be plated must be clean and free of grease;

2 the object to be plated should be rotated to give an even coating;

3 the electric current must not be too large or the 'coating' will form too rapidly and will flake off;

4 the temperature and concentration of the electrolyte must be carefully controlled, otherwise the 'coating' will be deposited too rapidly or too slowly.

The metal coating is deposited on the cathode and so the object to be plated must be the cathode during electrolysis. The electrolyte must contain a compound of the metal which forms the coating.

Figure 1 shows a nickel alloy fork being electroplated with silver. Notice that the fork is attached to a rotatable arm and that the electrolyte contains silver nitrate ($AgNO_3$).

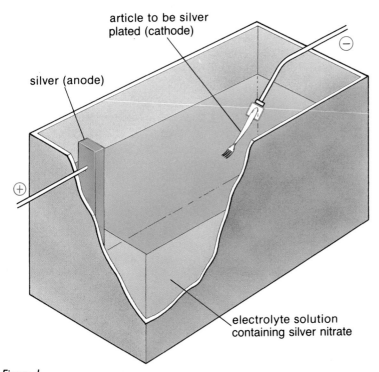

article to be silver plated (cathode)

silver (anode)

electrolyte solution containing silver nitrate

Figure 1
A fork being electroplated with silver

During electrolysis, silver ions (Ag^+) in the electrolyte are attracted to the cathode (the fork). Here they gain electrons and form a deposit of silver:

Cathode (−) $\qquad Ag^+ + e^- \rightarrow Ag$

The anode is a piece of silver. Nitrate ions (NO_3^-) in the electrolyte are attracted to the anode, but they are not discharged. Instead, silver atoms in the anode give up electrons to the anode and go into solution as Ag^+ ions.

Anode (+) $\qquad Ag \rightarrow Ag^+ + e^-$

Questions

1 What precautions should be taken to obtain a good coating of metal during electroplating?

2 (a) The object to be electroplated is made the cathode in an electrolytic cell. Why is this?
(b) Suppose you want to nickel plate an article. What substance would you choose for (i) the anode; (ii) the electrolyte?

3 Bath taps are electroplated with nickel and then with chromium.
(a) Why is the steel plated with a layer of nickel before chromium?
(b) Give two reasons for electroplating steel.

4 Some cutlery is stamped 'EPNS'. This stands for electroplated nickel silver. How is this plating done?

5 Articles of jewellery (e.g. bracelets) are often electroplated with gold, silver or copper. Why?

Section D: Activities

1 Recycling aluminium

Aluminium is obtained from bauxite which is impure aluminium oxide. If we continue to use aluminium at our present rate, the bauxite will only last another 30 years. This means that we should think about saving our reserves of aluminium. There are two ways in which most of us waste aluminium.

- milk bottle tops
- aluminium foil

In this activity, you will try to estimate

- how much aluminium we use each year
- how much we could save by recycling aluminium.

How much aluminium do we use each year?

1 Estimate the number of milk bottle tops your family uses (on average) each week.
2 Estimate the area of aluminium foil (in square centimetres) that your family uses each week for cooking, wrapping food, etc.
3 Use your answer to question **1** to work out the mass of aluminium which your family uses each week in milk bottle tops. (Assume that one milk bottle top weighs 0.25 g.)
4 Use your answer to question **2** to work out the mass of aluminium your family uses each week in foil. (Assume 100 cm^2 of foil weighs 0.4 g.)
5 Use your answers in questions **3** and **4** to calculate the total mass of aluminium (milk bottle tops plus foil) which your family uses in one week.
6 How much aluminium does your family use in one year?
7 Estimate the total mass of aluminium which we use in the UK in one year. (Assume that the population of the UK is 56 million.)

How much could we save by recycling aluminium?

8 Use your answer to question **7** to calculate the *total cost* of the aluminium which we use for milk bottle tops and foil each year in the UK. (Aluminium costs £800 per tonne from the factory. 1 tonne = 1000 kg.)
9 If possible, collect answers to question **8** from others in the class and calculate an average value.
10 Would we save the amount of money estimated in questions **8** and **9** if we recycled aluminium milk bottle tops and foil? Explain your answer.
11 About 35% of the aluminium used in the UK is recycled. What do you think are the difficulties in recycling more aluminium?
12 Which forms of aluminium are easiest to recycle? (Remember that aluminium has many other uses besides milk bottle tops and foil.)
13 What other common household materials could be recycled besides aluminium?
14 Why is it important to recycle resources, like aluminium?
15 What properties of aluminium make it particularly useful for (i) aircraft construction, (ii) greenhouse frames, (iii) cooking foil?

Recycling aluminium cans

2 | How much electricity is needed to deposit 1 mole of copper?

The apparatus below can be used to find the amount of electric charge required to deposit 1 mole of copper (63.5 g) on the cathode during electrolysis.

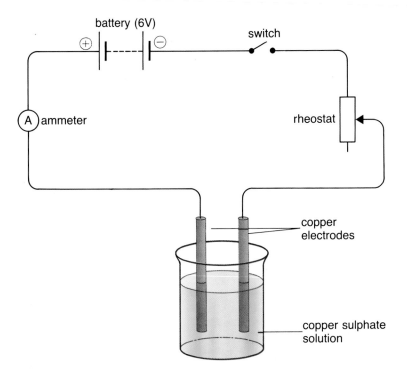

Here are the results of one experiment:

Mass of copper cathode before electrolysis = 51.23 g
Mass of copper cathode after electrolysis = 51.36 g
Time of electrolysis   = 22.5 minutes
Current   = 0.30 A

1 Why is the variable resistor (rheostat) used in the circuit?
2 Why must the cathode be clean and dry when it is weighed before electrolysis?
3 Why is the cathode washed in distilled water and then acetone after electrolysis?
4 What weight of copper was deposited during the experiment?
5 How much electricity (electric charge) passed during the experiment?
6 How much electricity (electric charge) is needed to deposit 1 mole (63.5 g) of copper?
7 A similar experiment to the last one was carried out using silver electrodes in silver nitrate solution. 0.45 g of silver was deposited in 22.5 minutes using a current of 0.30 A. Calculate the amount of charge (electricity) needed to deposit 1 mole of silver (108 g).
8 What do these experiments suggest about the relative size of the charge on one silver ion compared to that on one copper ion?

Section D: Study Questions

1 When aluminium is exposed to the air it becomes coated very quickly with a thin layer of oxide about 10^{-6} cm thick. This layer does not flake off, nor does it increase in thickness on standing. In order to protect the aluminium even more than its natural oxide layer does, it is possible to thicken the layer to 10^{-3} cm by a process called anodizing. The aluminium is first degreased and then anodized by making it the anode during the electrolysis of sulphuric acid. The oxygen released at the anode combines with the aluminium and increases the thickness of the oxide layer.

   (a) Why is anodizing useful?
   (b) Why does aluminium not corrode away like iron, even though aluminium is coated very quickly with a layer of oxide?
   (c) Write a word equation for the reaction which takes place when aluminium is exposed to the air.
   (d) Carbon tetrachloride (tetrachloromethane) can be used to degrease aluminium before anodizing.
Name one other liquid which would be a suitable degreasing agent.
   (e) Why is water not used to degrease aluminium?
   (f) Why is it necessary to degrease the aluminium?
   (g) How many times thicker is the oxide coating after anodizing?
   (h) Why is anodized aluminium especially useful as a building material?

2 This question concerns the properties of substances *A* to *E*.

	Electrical conductivity (of the pure substance) at room temperature	Solubility in water	Properties of the solution in water
A	Conductor	Reacts with water	An alkaline solution which conducts electricity
B	Conductor	Insoluble	—
C	Non-conductor	Soluble	A neutral solution which does not conduct
D	Non-conductor	Soluble	A neutral solution which conducts electricity
E	Non-conductor	Insoluble	—

Select, from *A* to *E*, the correct substance for:
   (a) the wiring for an electric light
   (b) the insulator for overhead electricity cables
   (c) a quick acting fertilizer
   (d) glass
   (e) sugar
   (f) calcium.

ULEAC

3 A solution of copper sulphate is electrolysed using copper electrodes.
   (a) Draw a clearly labelled diagram of the circuit.
   (b) What happens during electrolysis
   (i) at the anode
   (ii) at the cathode
   (iii) to the solution?
   (c) Mention two practical uses of this type of electrolysis.
   (d) If the two copper electrodes are replaced by two platinum electrodes, state what happens now
   (i) at the anode
   (ii) at the cathode
   (iii) to the solution.

4

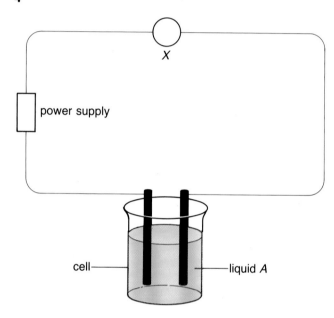

Liquid *A* is to be electrolysed.
   (a) What type of power supply is suitable for this purpose?   (1)
   (b) What piece of apparatus could be connected at *X* to show that liquid *A* is a conductor?   (1)
   (c) Give the name of the charged particles which conduct the electric current in
   (i) the connecting wires
   (ii) the electrolyte.   (2½)
   (d) A metal kettle is to be coated with copper using electrolysis. State what you would use in the cell as the
   (i) anode
   (ii) cathode
   (iii) electrolyte.   (3)

ULEAC

SECTION E
Patterns and Properties

The periodic table is like a jigsaw. The pieces must be arranged correctly to give a pattern or a picture. In the periodic table, elements are arranged in order of relative atomic mass to give a pattern in properties

1 Patterns of Elements

Why are graphite strips used around the nose cones of aircraft and rockets like this?

One useful way of classifying elements is as metals and non-metals (see section A, unit 4). Unfortunately, it is not easy to classify some elements in this way. Take, for example, graphite and silicon. These two elements have high melting points and high boiling points (like metals) but they have low densities (like non-metals). They conduct electricity better than non-metals but not as well as metals. Elements with some properties like metals and other properties like non-metals are called **metalloids**.

Because of this limitation in classifying elements neatly as metals and non-metals, chemists looked for patterns in the properties and reactions of smaller groups of elements.

Mendeléev's periodic table

During the nineteenth century, several chemists tried to find some pattern in the properties of elements.

The most successful of these attempts was made by the Russian chemist Dmitri Mendeléev in 1869.

Mendeléev wrote down all the known elements in order of their relative atomic masses. He then arranged the elements in horizontal rows so that elements with similar properties appeared in the same vertical column. Because of the periodic repetition of elements with similar properties, Mendeléev called his arrangement a **periodic table**.

Figure 1 shows part of Mendeléev's periodic table. Notice that elements with similar properties, such as sodium and potassium, fall in the same vertical column. Which other pairs or trios of similar elements appear in the same vertical column of Mendeléev's table?

> In the periodic table,
> * The vertical columns of similar elements are called **groups**.
> * The horizontal rows are called **periods**.

Mendeléev had some brilliant and successful ideas in connection with his periodic table.

* He left gaps in his table so that similar elements were in the same vertical group. Four of these gaps are shown as asterisks in Figure 1.

* He suggested that elements would be discovered to fill the gaps. The four gaps in Figure 1 are now filled by the elements scandium, gallium, germanium and technetium.

* He predicted the properties of the missing elements from the properties of elements above and below them in his table.

Dmitri Mendeléev was the first chemist to successfully arrange the elements into a pattern linking their properties and relative atomic masses

Within 15 years of Mendeléev's predictions, three of the missing elements had been discovered. Their properties were very similar to his predictions.

The success of Mendeléev's predictions showed that his ideas were probably correct. His periodic table was quickly accepted as an important summary of the properties of elements.

					GROUP				
	I	**II**	**III**	**IV**	**V**	**VI**	**VII**	**VIII**
Period 1	H							
Period 2	Li	Be	B	C	N	O	F	
Period 3	Na	Mg	Al	Si	P	S	Cl	
Period 4	K	Ca	*	Ti	V	Cr	Mn	Fe Co Ni
	Cu	Zn	*	*	As	Se	Br	
Period 5	Rb	Sr	Y	Zr	Nb	Mo	*	Ru Rh Pd
	Ag	Cd	In	Sn	Sb	Te	I	

Figure 1

Questions

1 (a) Why did scientists start to use the term 'metalloid'?
(b) Name three metalloids.

2 Look at Table 1. This shows the predictions which Mendeléev made in 1871 for the element in period 4 below silicon. This element was discovered in 1886.

(a) What is the name for this element?
(b) Use a data book to check Mendeléev's predictions.
(c) How accurate were Mendeléev's predictions?

3 (a) Use a data book to look up the properties of aluminium (Al) and yttrium (Y).
(b) Make predictions for the element between Al and Y in Mendeléev's periodic table.
(c) Use a modern periodic table to identify the element between Al and Y.
(d) Check your predictions for this element against its actual properties.

Mendeléev's predictions for the element below silicon

1 grey metal
2 density 5.5 g cm^{-3}
3 relative atomic mass = average of relative atomic mass of Si and Sn
$$= \frac{28.1 + 118.7}{2} = 73.4$$
4 melting point higher than that of tin—perhaps about 800° C
5 formula of oxide will be XO_2.
Density of $XO_2 = 4.7$ g cm^{-3}
6 the oxide XO may also exist

Table 1

2 Modern Periodic Tables

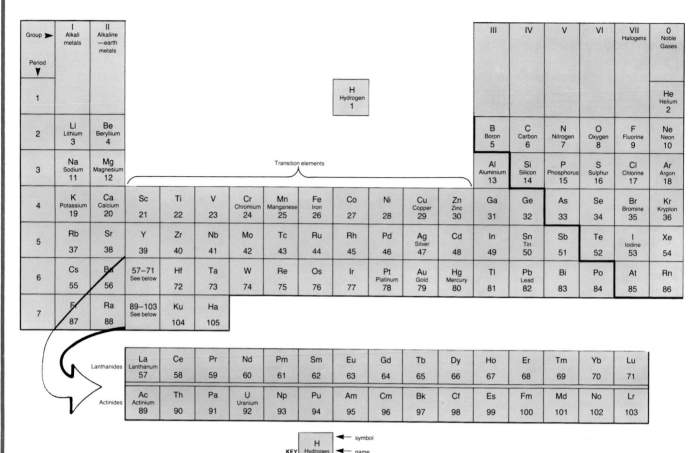

Figure 1
The modern periodic table

All modern periodic tables are based on the one proposed by Mendeléev in 1869. A modern periodic table is shown in figure 1. The elements are numbered along each period, starting with period 1, then period 2, etc. The number given to each element is called its **atomic number**. Thus, hydrogen has an atomic number of 1, helium 2, lithium 3, etc. You will learn more about atomic numbers in section K.

There are several points to note about the modern periodic table.

1 The most obvious difference between modern periodic tables and Mendeléev's is the position of the **transition elements**. These have been taken out of the simple groups and placed between group II and group III. Period 4 is the first to contain a series of transition elements. These include chromium, iron, nickel, copper and zinc.

2 Some groups have names as well as numbers. These are summarized in the table (left).

3 Metals are clearly separated from non-metals. The 20 or so non-metals are packed into the top right-hand corner above the thick stepped line in figure 1. Some elements close to the steps are metalloids. These elements have some properties like metals and some properties like non-metals.

Group number	Group name
I	alkali metals
II	alkaline-earth metals
VII	halogens
O	noble (inert) gases

The names of groups in the periodic table

4 Apart from the noble gases, the most reactive elements are near the left- and right-hand sides of the periodic table. The least reactive elements are in the centre. Sodium and potassium, two very reactive metals, are at the left-hand side. The next most reactive metals, like calcium and magnesium, are in group II, whereas less reactive metals (like iron and copper) are in the centre of the table. Carbon and silicon, unreactive non-metals, are in the centre of the periodic table. Sulphur and oxygen, which are nearer the right-hand edge, are more reactive. Fluorine and chlorine, the most reactive non-metals, are very close to the right-hand edge.

Group

Period	I	II												III	IV	V	VI	VII	O
Period 1						H													He
Period 2	Li	Be												B	C	N	O	F	Ne
Period 3	Na	Mg												Al	Si	P	S	Cl	Ar
Period 4	K	Ca	Sc	Ti	V	Cr	Mn	Fe	Co	Ni	Cu	Zn		Ga	Ge	As	Se	Br	Kr
Period 5	Rb	Sr	Y	Zr	Nb	Mo	Tc	Ru	Rh	Pd	Ag	Cd		In	Sn	Sb	Te	I	Xe
Period 6	Cs	Ba	La	Hf	Ta	W	Re	Os	Ir	Pt	Au	Hg		Tl	Pb	Bi	Po	At	Rn
Period 7	Fr	Ra	Ac	Ku	Ha														

Figure 2
Blocks of similar elements in the periodic table

Questions

1 In modern periodic tables, you can pick out 5 blocks of elements with similar properties. These blocks are shown in different colours in figure 2.

(a) The five blocks of similar elements are called non-metals, noble gases, poor metals, reactive metals and transition metals. Which name belongs to which coloured block in figure 2?

(b) Which groups make up the reactive metals?

(c) In the transition metals, the elements resemble each other across the series as well as down the groups. Pick out two sets of three elements to illustrate these similarities.

(d) Some poor metals have properties like non-metals. Give two examples of this.

(e) The noble gases are very unreactive. The first noble gas compound was not made until 1962. Today, several compounds of them are known.

(i) The noble gases were once called 'inert gases'. Why was this?

(ii) Why do you think their name was changed to 'noble gases'?

(iii) Why are there no noble gases in Mendeléev's periodic table?

2 Draw an outline of the periodic table similar to figure 2. On your outline, indicate where you would find (i) metals, (ii) non-metals, (iii) metalloids, (iv) elements with atomic numbers 11 to 18 inclusive, (v) the alkaline-earth metals, (vi) the most reactive metal, (vii) the most reactive non-metal, (viii) elements that might be used as disinfectants, (ix) gases, (x) magnetic elements, (xi) elements used in jewellery, (xii) elements with the highest densities.

3 Suppose you are Mendeléev. The year is 1869. You are just about to announce the discovery of your periodic table by writing a letter to the President of the Russian Academy of Sciences. Write down what you would say.

3 The Alkali Metals

Group 1

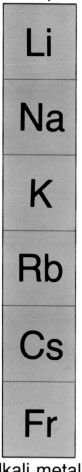

Alkali metals

Figure 1

The elements in Group I are called alkali metals because they react with water to form alkaline solutions. Lithium, sodium and potassium are the best known alkali metals. The other elements in Group I are rubidium, caesium and francium (figure 1).

The alkali metals are so reactive that they must be stored under oil to protect them from oxygen and water vapour in the air. Some properties of lithium, sodium and potassium are summarized in Table 1.

Property	**Character**
Appearance	Shiny but quickly form a dull layer of oxide
Strength	Soft metals—easily cut with a knife
M.p. and b.pt.	Low compared with other metals
Density	Less than 1.0 g cm^{-3}—float on water
Reaction with air	Burn vigorously forming white oxides, e.g. sodium + oxygen → sodium oxide $4Na \quad + \quad O_2 \quad \to \quad 2Na_2O$
Reaction with cold water	Lithium reacts steadily, sodium vigorously, potassium violently. The products are hydrogen and an alkaline solution of the metal hydroxide, e.g. sodium + water → sodium hydroxide + hydrogen $2Na \quad + \quad 2H_2O \to \quad 2NaOH \quad + \quad H_2$
Valency (combining power)	All have a valency of one Oxides are Li_2O, Na_2O, K_2O Ions are Li^+, Na^+, K^+ Salts are white unless the anion is coloured

Table 1: Similarities of lithium, sodium and potassium

Notice how the elements react more vigorously with water from lithium to potassium. This is due to an increase in metallic character down the group. The ions of all alkali metals have a charge of +1. So all their compounds will have similar formulas.

Notice also in table 1 that alkali metals have some unusual properties for metals

- Their melting points and boiling points are unusually low

- Their densities are so low that they float on water

- They are soft enough to be cut with a knife

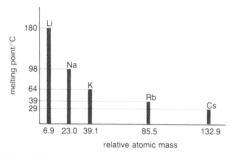

Figure 2

Some of the physical properties of lithium, sodium and potassium are given in Table 2.

Element	Relative atomic mass	Melting point/°C	Boiling point/°C	Density /g cm^{-3}
Lithium	6.9	180	1330	0.53
Sodium	23.0	98	892	0.97
Potassium	39.1	64	760	0.86

Table 2: The physical properties of lithium, sodium and potassium

The large cylinder carries liquid sodium which is used as a coolant for nuclear reactors

(1) These properties illustrate another important feature of the periodic table—*although the elements within a group are similar, there is a steady change in property from one element to the next.* The graph in figure 2 shows the steady change in the melting points of the alkali metals very neatly.

Once we know the general properties of the elements in a group, and how these properties vary from one element to the next, we can predict the properties of other elements in the group.

Rock salt being processed above ground

The uses of sodium compounds

Most sodium compounds are made from sodium chloride. In hot countries, impure sodium chloride is left when sea water is allowed to evaporate. Sodium chloride also occurs in salt beds beneath the Earth's surface. The sodium chloride can be obtained from these salt beds by *solution mining.* This involves piping hot water down to the salt bed to dissolve the salt. Concentrated salt solution is then pumped to the surface. Impure salt is used for de-icing roads. Pure salt is used in cooking. Sodium and chlorine can be produced by electrolysing molten salt (section D, unit 4). Liquid sodium is used as a coolant in fast nuclear reactors and sodium vapour provides the yellow glow in street lamps.

Questions

1 Write word equations and then balanced equations with symbols for the reaction of potassium with (i) oxygen, (ii) water, (iii) chlorine.

2 Rubidium (Rb) is in Group I below potassium. Use the information in tables 1 and 2 to predict (i) its melting point, (ii) its boiling point, (iii) its density relative to that of water, (iv) its relative atomic mass, (v) the symbol of its ion, (vi) the formula of its oxide and chloride, (vii) its reaction with water, (viii) how it burns in air.

3 Find out about (i) the manufacture of sodium, (ii) the use of liquid sodium as a coolant, (iii) the use of sodium vapour in street lamps, (iv) the manufacture of other sodium compounds from sodium chloride.

4 Draw a poster or diagram illustrating or summarizing the manufacture and uses of sodium and its compounds.

5 A metal *M* is in Group III of the periodic table. Its chloride has a formula mass of 176.5.
(a) Write the formula of the chloride of the metal.
(b) Calculate the atomic mass of the metal. (Chlorine has relative atomic mass = 35.5.)
(c) Which element is M?

4 The Transition Metals

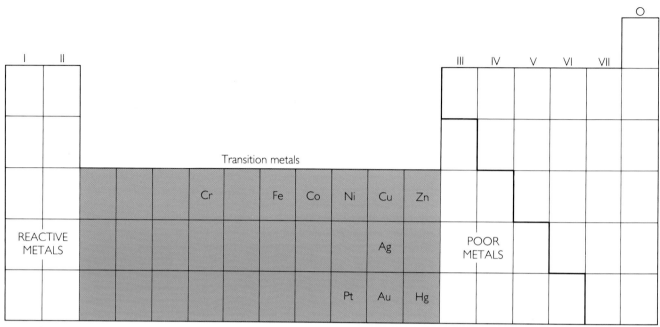

Figure 1

Figure 1 shows some of the common transition metals in the periodic table. The transition metals separate the reactive metals in Groups I and II from the poor metals in the triangle below the steps separating metals and non-metals.

> *The transition metals have very similar properties. Unlike other parts of the periodic table, there are similarities in the transition metals across the periods as well as down the groups.*

The most important transition metals are iron and copper. Iron is the most widely used metal. More than 700 million tonnes of it are manufactured every year throughout the world. It is used as steel in girders and supports for bridges and buildings, in vehicles, in engines and in tools.

After iron and aluminium, copper is the third most widely used metal. About 9 million tonnes are manufactured each year. Copper is a good conductor of heat and electricity. It is also malleable (it can be made into different shapes) and ductile (it can be drawn into wires). Because of these properties, copper is used in electrical wires and cables and in hot water pipes and radiators. The uses of copper are increased by *alloying* it with other metals. Alloying copper with zinc produces brass which is harder and stronger than pure copper. Alloying copper with tin produces bronze. This is stronger and easier to cast into moulds than pure copper. There is more about alloys in section F.

General properties of transition metals

Some properties of iron and copper are shown in the table (above right). Compare these with the properties of alkali metals in tables 1 and 2 in the last unit.

This statue has been made from copper. Copper is used because it is easy to shape and its colour gives an attractive finish

Element	Melting point /°C	Boiling point /°C	Density /g cm^{-3}	Reaction with water	Formulas of oxides	Symbols of ions	Salts
Iron	1540	3000	7.9	Does not react with pure water. Reacts slowly with steam.	FeO Fe_2O_3	Fe^{2+} Fe^{3+}	Fe^{2+} salts are green. Fe^{3+} salts are yellow or brown.
Copper	1080	2600	8.9	No reaction with water or steam.	Cu_2O CuO	Cu^+ Cu^{2+}	Cu^{2+} salts are blue or green.

Properties of iron and copper

The information in the table above, illustrates some typical properties of transition metals.

1 High melting points and boiling points—much higher than alkali metals.

2 High densities—much higher than alkali metals.

3 Hard strong metals (high tensile strength), unlike the soft alkali metals.

Iron is so strong that it is used as steel in girders and supports such as those in this oil rig under construction

4 Fairly unreactive with water, unlike alkali metals. None of the transition metals react with cold water, but a few of them react slowly with steam, like iron.

5 More than one valency. Most of the transition metals can have more than one valency: Iron forms Fe^{2+} ions (valency 2) and Fe^{3+} ions (valency 3); Copper forms Cu^+ ions (valency 1) and Cu^{2+} ions (valency 2). Alkali metals only form ions of 1 + charge (valency 1).

6 Coloured compounds. Transition metals usually have coloured compounds with coloured solutions. In contrast, alkali metals have white salts with colourless solutions.

7 Catalytic properties. Transition metals and their compounds can act as **catalysts** (see section K). Iron or iron(III) oxide (Fe_2O_3) is used as a catalyst in the *Haber process* to manufacture ammonia (NH_3).

Questions

1 What are the transition metals?
2 Make a list of the characteristic properties of transition metals.
3 Make a table contrasting the properties of transition metals with those of alkali metals.
4 Write equations with symbols for the following word equations:
  (a) iron + oxygen → iron(III) oxide
  (b) iron + chlorine → iron(III) chloride
  (c) iron + hydrogen chloride → iron(II) chloride + hydrogen.
5 What is (i) an alloy; (ii) brass; (iii) bronze?
6 (a) Make a list of the important uses of (i) iron; (ii) copper.
  (b) Explain these uses in terms of the properties of iron and copper.

5 The Halogens

	VI	VII	0
			He
	O	F	Ne
	S	Cl	Ar
		Br	Kr
		I	Xe
		At	Rn

Figure 1

Pale yellow cubic crystals of fluorite (CaF₂). Fluorite is a common source of fluorine. The small black crystals are galena (PbS)

The elements in Group VII of the periodic table are called **halogens** (figure 1). They are a group of reactive non-metals. The common elements in the group are fluorine (F), chlorine (Cl), bromine (Br) and iodine (I). The final element, astatine (At), does not occur naturally. It is an unstable radioactive element which was first synthesized in 1940.

The halogens are so reactive that they never occur free in nature. They occur combined with metals in salts such as sodium chloride (NaCl), calcium fluoride (CaF$_2$) and magnesium bromide (MgBr$_2$). This gave rise to the name for the group as halogens which means 'salt-formers'.

Sources of the halogens

The halogens occur in compounds with metals as negative ions: fluoride (F$^-$), chloride (Cl$^-$), bromide (Br$^-$) and iodide (I$^-$). The most widespread compound containing fluorine is fluorspar or fluorite (CaF$_2$). The commonest chlorine compound is sodium chloride (NaCl) which occurs in sea-water and in rock salt. Each kilogram of sea-water contains about 30 g of sodium chloride. Sea-water also contains small amounts of bromides and traces of iodides. Certain seaweeds also contain iodine but the main source of iodine is sodium iodate (NaIO$_3$). This is found in Chile mixed with larger quantities of sodium nitrate (NaNO$_3$).

Laminaria is a type of seaweed which contains iodine. You can often find laminaria at low tide on rocky beaches around Britain. Some types of seaweed are edible

Patterns in physical properties

The table opposite lists some physical properties of chlorine, bromine and iodine.

Look at the table and decide how the following properties of the halogens change as their relative atomic mass increases?

1 state at room temperature;
2 colour of vapour;
3 melting points (see figure 2);
4 boiling points (see figure 2).

Element	Relative Atomic Mass	Colour and state at room temperature	Colour of vapour	Structure	M.pt. /°C	B.pt. /°C
Chlorine	35.5	Pale green gas	Pale green	Cl_2 molecules	−101	−35
Bromine	79.9	Red brown liquid	Orange	Br_2 molecules	−7	58
Iodine	126.9	Dark grey solid	Purple	I_2 molecules	114	183

Physical properties of chlorine, bromine and iodine

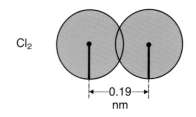

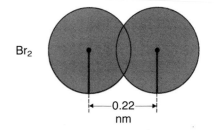

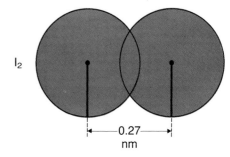

Figure 3

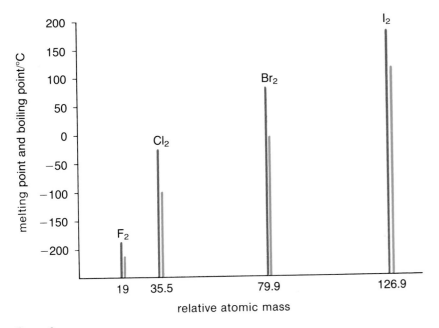

Figure 2
Melting points and boiling points of the halogens

The halogens form diatomic molecules, F_2, Cl_2, Br_2 and I_2. The relative sizes of these molecules (figure 3) explain the changes in volatility (i.e. melting points and boiling points) with increasing relative atomic mass.

The larger the molecule, the harder it is to break up the orderly arrangement within the solid to form the liquid. It is also more difficult to separate larger molecules and form the gas. This is because there are stronger intermolecular forces between the larger molecules than the smaller ones. Thus, more energy is needed to melt and to boil the larger molecules. Hence the melting points and boiling points increase with relative molecular mass. F_2 and Cl_2, the smallest halogen molecules, are gases at room temperature. Br_2 is a liquid and I_2 is a solid.

Questions

1 What are the halogens?
2 What are the most important sources of (i) chlorine; (ii) bromine; (iii) iodine?
3 List the characteristic physical properties of the halogens.
4 Use the information in this unit to predict the following properties of fluorine and astatine:
colour; state at room temperature; structure; melting point; boiling point.
5 Explain the meaning of the following terms which are used in this unit:
unstable radioactive element, synthesized, diatomic molecule, volatility, intermolecular forces.

6 Reactions of Chlorine and the Halogens

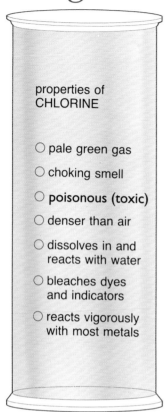

properties of
CHLORINE

○ pale green gas

○ choking smell

○ **poisonous (toxic)**

○ denser than air

○ dissolves in and
   reacts with water

○ bleaches dyes
   and indicators

○ reacts vigorously
   with most metals

Figure 1

Household bleaches, like Domestos, usually
contain chlorine itself and chlorine compounds.
As these substances are so reactive they must
be handled with great care

Chlorine

Chlorine is the most important element in Group VII. Chlorine and its compounds have many more uses than the other halogens and we shall look at these in the next unit. The important properties of chlorine are summarized in figure 1. *Remember that chlorine has a choking smell and that it is poisonous (toxic). Always treat it with care. Handle in a fume cupboard, except in very small quantities.*

Tests for chlorine

The best test for chlorine is to show that it will bleach moist litmus paper. If blue litmus paper is used, the paper first turns red (because chlorine reacts with water to form an acidic solution) and then goes white.

Reactions with water

When chlorine is bubbled into water, it reacts to form a mixture of hydrochloric acid and hypochlorous acids:

chlorine + water → hydrochloric acid + hypochlorous acid
$$Cl_2 \ + \ H_2O \ \rightarrow \ HCl \ + \ HClO$$

The solution produced is very acidic. It is also a strong bleach which quickly turns litmus paper or universal indicator paper white. Bromine reacts less easily, but in a similar way to chlorine. The solution produced is orange red and acidic. It bleaches less rapidly than chlorine water.

Iodine dissolves in water only very slightly. The solution is a pale yellow colour. It is only slightly acidic and it bleaches very slowly.

Household bleaches, like Domestos, usually contain chlorine itself and chlorine compounds. These compounds are so reactive that they remove coloured stains by reacting with them to make colourless substances.

Reactions with iron

The reaction between chlorine and iron can be studied using the apparatus in figure 2. The iron wool is heated to start the reaction and then the Bunsen is removed. The iron glows as it reacts with the chlorine. The product is brown iron(III) chloride which sublimes along the tube:

iron + chlorine → iron(III) chloride
$$2Fe \ + \ 3Cl_2 \ \rightarrow \ 2FeCl_3$$

If you try these experiments
wear eye protection
and use a fume
cupboard

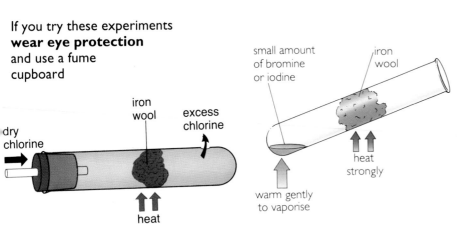

Figure 2

Figure 3

Industrial bleaches containing chlorine and chlorine compounds are used to bleach paper and cloth. If chlorine-based products are allowed to leak into the environment, they can cause serious pollution

Figure 3 shows how we can investigate the reactions of iron with bromine or iodine. When iron is heated in bromine vapour, the reaction is less vigorous than that with chlorine. The iron glows but it must be heated all the time. The product is iron(III) bromide. When the experiment is repeated with iodine, the iron reacts very slowly to form iron(II) iodide.

Comparing reactivities

The reactions of halogens with water, with iron and as bleaching agents shows their relative reactivities very clearly. Chlorine is more reactive than bromine, and bromine is more reactive than iodine. Notice that *the halogens get less reactive as their relative atomic mass increases.* This is opposite to the trend in Group I. The alkali metals get *more* reactive with increasing relative atomic mass.

Chlorine as an oxidizing agent

When chlorine reacts with metals, it gains electrons from them, forming negative chloride ions. For example,

$$2Na + Cl_2 \rightarrow 2Na^+Cl^-$$

This is like the reaction of oxygen with metals to form oxide ions. For example,

$$2Mg + O_2 \rightarrow 2Mg^{2+}O^{2-}$$

The oxygen acts as an oxidizing agent by accepting electrons from the metal. In the same way, chlorine can be regarded as an oxidizing agent since it accepts electrons from the metal. Chlorine also acts as an oxidizing agent when it:

- bleaches dyes and indicators;
- reacts with green solutions of Fe^{2+} ions forming pale yellow Fe^{3+} ions:

$$2Fe^{2+} \rightarrow 2Fe^{3+} + 2e^-$$

$$Cl_2 + 2e^- \rightarrow 2Cl^-$$

Questions

1 Describe experiments which show the relative reactivity of bromine, chlorine and iodine. Write word equations and balanced equations for the reactions you describe.

2 Arrange the following pairs of elements in order of decreasing vigour of reaction under the same conditions.
lithium and iodine, potassium and chlorine, potassium and fluorine, sodium and chlorine, sodium and iodine

3 The following poem by Vernon Newton is called 'Mistress Fluorine'.
Fervid Fluorine, though just nine,
Knows her aim in life: combine!
In fact, of things that like to mingle,
None's less likely to stay single.
  (a) Use a dictionary to find the meaning of 'fervid' (line 1) and 'mingle' (line 3).
  (b) Why is fluorine described as 'fervid'?
  (c) Why is fluorine described as 'just nine'? (line 1)
  (d) What does the poem say about the properties of fluorine?
  (e) Try to write a short poem about any other element.

---

7 The Uses of Chlorine

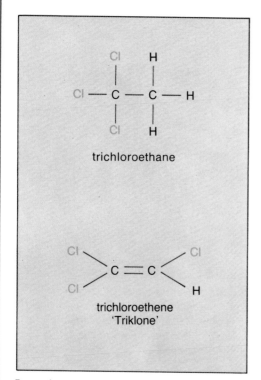

Figure 1
Solvents containing chlorine

Chlorine and its compounds have far more uses than the other halogens. They are important in industry, agriculture, medicine and the home. Two important uses of the element are in water sterilization and in bleaches. Chlorine is added in small quantities to our water supplies. The chlorine kills bacteria in the water, but it does not affect humans or animals. Larger quantities of chlorine are added to swimming baths. The use of chlorine in domestic bleaches was discussed in unit 6 of this section.

Uses of chlorine compounds

● **Hydrochloric acid.** Large quantities of chlorine are used to make hydrochloric acid. Chlorine is first burnt in hydrogen to form hydrogen chloride:

$$\text{hydrogen} + \text{chlorine} \rightarrow \text{hydrogen chloride}$$
$$H_2 + Cl_2 \rightarrow 2HCl$$

The hydrogen chloride is then dissolved in water to produce hydrochloric acid. This is the cheapest industrial acid. It is used to clean the rust from steel sheets before galvanizing and to produce ammonium chloride.

● **Solvents.** Many of the solvents used to remove oil, fat and grease from clothing and machinery are compunds containing chlorine (figure 1). Tetrachloroethene is used in dry-cleaning clothes and trichloroethene ('Triklone') is used to degrease machinery. The 'thinner' (solvent) for Tippex is trichloroethane.

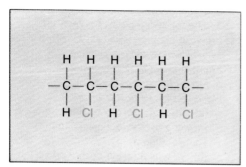

Figure 2
A section of PVC polymer

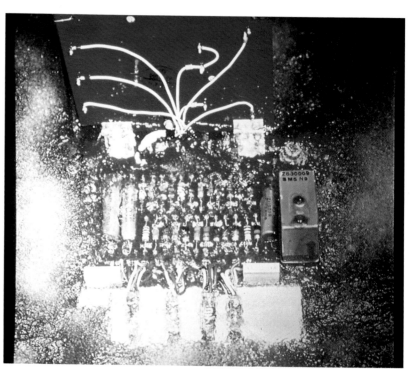

Circuit boards being degreased and cleaned by dipping in Arklone, a chlorine-containing solvent

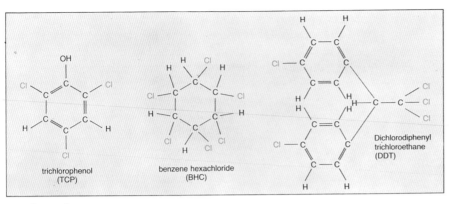

Figure 3
Disinfectants and insecticides containing chlorine

trichlorophenol (TCP)

benzene hexachloride (BHC)

Dichlorodiphenyl trichloroethane (DDT)

All these products are made from PVC

- **Plastics.** One of the most important plastics is PVC (polyvinyl chloride). The correct chemical name for this is polychloroethene (figure 2). PVC plastics are used for raincoats, coverings for tables and shelves ('Fablon'), floor tiles ('Vinyl' tiles), upholstery, records and electrical insulation.

- **Disinfectants.** Two of the most widely used general disinfectants are TCP (trichlorophenol—figure 3) and Dettol, which is a similar chlorine-containing compound. These are used in many homes for treating cuts, gargling and cleaning sinks and toilets.

- **Insecticides.** During the 1940s, DDT (dichlorodiphenyltrichloroethane—figure 3) was produced and used as an insecticide. It killed lice, flies and mosquitoes which were attacking crops and causing typhoid, dysentery and malaria.

During the 1960s, it became clear that DDT was getting into food chains and killing birds and animals. The use of DDT was therefore banned in some countries including the U.K. Its use as a pesticide has been replaced by various other insecticides such as BHC (figure 3).

Questions

1 (a) How is hydrochloric acid manufactured?
(b) What are its uses?

2 Read the following poem from *The Penguin Book of Limericks*.

*A mosquito was heard to complain
That a chemist had poisoned his brain;
The cause of his sorrow
Was 4,4-dichloro-
Diphenyltrichloroethane.*

(a) Why do you think 4,4-dichlorodiphenyltrichloroethane is called DDT?
(b) What elements does DDT contain?
(c) What is the molecular formula of DDT?
(d) What use of DDT does the poem illustrate?
(e) During the 1960s, the bodies of birds of prey were found to contain unusually large amounts of DDT. Why was this?
(f) Why has our use of DDT changed since the 1940s?
(g) Design a poster to show the dangers of using large amounts of DDT.

3 Use moleculear models to build the structures of the substances in figures 1, 2 and 3. How do your molecular models compare with the structures drawn in the figures?

DDT being used to delouse captured soldiers during the Korean War

8 The Noble Gases

Figure 1

Figure 1 shows the position of the noble gases in the periodic table. It also includes some other well-known non-metals. *The discovery of the noble gases showed the real value of the periodic table.* In 1868, an orange line was noticed in the spectrum of light from the Sun. This showed that the Sun's atmosphere contained an element that had not been found on the Earth. The element was named helium which comes from the Greek word *helios* meaning Sun.

In 1890, Raleigh and Ramsay prepared what they thought was pure nitrogen by removing oxygen, water vapour and carbon dioxide from fresh air. They found that the density of this gas differed by 0.5% from that of pure nitrogen obtained by decomposing nitrogen compounds. Further experiments on the impure nitrogen from the air showed that it contained a new element, argon, with a relative atomic mass of 40.

Later, Ramsay found helium on Earth and showed that it was very similar to argon. However, these two elements were very different from any of the other elements in the periodic table. This suggested that there was another group in the periodic table. The group was called Group O and a search began for the other four elements in it.

The fractional distillation of liquid air (section B, unit 1) led to the discovery of three of these elements (neon, krypton and xenon) by Ramsay in 1898. The remaining element, radon, was discovered in 1900 as a product from the breakdown of the radioactive element, radium.

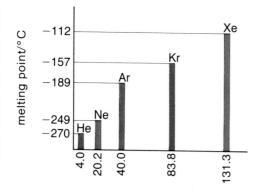

Sir William Ramsay helped in the discovery and identification of five of the noble gases

Element	Relative atomic mass	Melting point /°C	Boiling point /°C	Density at 20°C and atm. pressure /g dm^{-3}
Helium	4.0	−270	−269	0.17
Neon	20.2	−249	−246	0.83
Argon	40.0	−189	−186	1.7
Krypton	83.8	−157	−152	3.5
Xenon	131.3	−112	−108	5.5

Properties of the noble gases

Properties of the noble gases

The table lists some properties of the noble gases. They are all colourless gases at room temperature with low melting points and boiling points. As expected, their melting points, boiling points and densities show a steady change as the relative atomic mass increases. The graph in figure 2 shows the steady increase in melting point with relative atomic mass.

The noble gases all exist as separate single atoms (i.e. monatomic molecules). Until 1962, no compounds of the noble gases were known. Chemists thought they were completely unreactive. Hence they were called the *inert* gases. Nowadays, several compounds of them are known and the name *inert* has been replaced by *noble*.

Figure 2
A graph showing the increase in the melting points of the noble gases as their relative atomic masses increase

Obtaining the noble gases

Neon, argon, krypton and xenon are obtained industrially during the fractional distillation of liquid air. There are only minute traces of helium in air. Therefore, it is more economical to extract helium from the natural gas in oil wells.

Uses of the noble gases

Helium is used in meteorological balloons because of its low density and because it is non-flammable. Helium, mixed with oxygen, is also used as the gas breathed by divers.

Neon is used in neon lights, and argon and krypton are used in electric light bulbs. If there was a vacuum inside the bulbs, metal atoms would evaporate from the very, very hot tungsten filament. To reduce this evaporation and to prolong the life of the filament, the bulb is filled with an unreactive gas which cannot react with the hot tungsten filament.

Argon is also used during certain welding processes. The argon provides an inert atmosphere and prevents any reaction between the metals being welded and oxygen in the air.

The cylinders of this diving apparatus contain oxygen and helium

Helium being used to fill a weather balloon

Questions

1 (a) Which elements make up the noble gases?
(b) Why are they called *noble gases*?
(c) Why were they once called *inert gases*?

2 Use the values in the table to plot a graph of the boiling points of the noble gases (vertically) against their relative atomic masses (horizontally).
(a) How do the boiling points vary with relative atomic mass?
(b) Explain the pattern shown by the graph.
(c) Use the graph to predict the boiling point of radon (RAM = 222).

3 Suppose that liquid air contains oxygen (b.pt. $-183°$ C), nitrogen (b.pt. $-196°$ C), water (b.pt. $100°$ C), neon, argon, krypton and xenon. If the liquid air is fractionally distilled, what is the order in which the constituents will boil away from the liquid air?

4 Make a list summarizing the properties of the noble gases.

Section E: Activities

How is unleaded petrol treated to reduce the possibility of knocking?

1 The discovery of anti-knock compounds

Between 1910 and 1920, motor car engines became more complicated. This resulted in 'knocking' or 'pinking'. The mixture of petrol vapour and air in a car cylinder should *not* explode until the piston is at the top of the cylinder. Knocking occurs when the mixture explodes too early. This causes the engine to work in fits and starts.

In 1916, an engineer called Thomas Midgley got a job with an American oil company. Midgley was asked to investigate how knocking could be prevented. He tried various substances which were soluble in petrol to see if they reduced knocking. Iodine and iodine compounds were found to reduce knocking, but they were rather expensive.

So Midgley tried other halogens and their compounds (table 1). Chlorine, bromine and their compounds did not work as well as iodine and its compounds.

Next, Midgley tried elements and their compounds from Group VI. Selenium compounds reduced knocking better than iodine compounds. Tellurium compounds reduced knocking even better than selenium compounds. Unfortunately, tellurium compounds were extremely poisonous. Midgley studied the patterns in anti-knock properties very carefully. He deduced that the best anti-knock substance would be a compound of lead. This prediction turned out to be correct. Since then, tetraethyllead(IV) $(Pb(C_2H_5)_4)$ has been added to most petrol to prevent knocking.

Group IV	Group V	Group VI	Group VII
C	N	O	F
Si	P	S	Cl
Ge	As	Se	Br
Sn	Sb	Te	I
Pb	Bi	Po	At

Table 1: groups IV, V, VI, and VII in the periodic table

1 Iodine and its compounds have not been used commercially as anti-knock agents. Why is this?

2 Tellurium and its compounds have not been used commercially as anti-knock agents. Why is this?

3 Why was it pointless to test compounds of elements in Groups I, II and III (e.g. sodium chloride) as anti-knock agents?

4 How do anti-knock properties change (i) down a group, (ii) across a period?

5 Why did Midgley predict that the best anti-knock substance would be a compound of lead?

6 Why do you think Midgley never tested compounds of polonium and astatine?

7 Why is unleaded petrol being used more and more nowadays?

2 Ecology or economics—a mining enquiry

The facts

Geologists working for the mining company Zinc UK have discovered large deposits of zinc ore in an attractive coastal area of Southshire (figure 1). The ore contains up to 24% zinc which is very rich indeed. There is about 50 million tonnes of ore, 40 metres below the surface, in an area of land owned by an absentee landlord. This could be mined at a rate of about 2 million tonnes per year. The project would create about 200 new jobs in Southshire.

The issues

(i) *Environmental.* The area is regarded as one of outstanding natural beauty. It supports many wild plants and animals including deer and otter. It is a tourist attraction for visitors to the small seaside resorts along the coast. Most jobs in the area are related to dairy farming and tourism. Ten miles from the proposed mine there is a new housing estate at Whitford with room for expansion. Whitford has a few shops, a coffee bar, a pub and a primary school. The seaside resorts are small because of the poor road links and because there is no railway. The mining company have, however, promised to build a major road into the area if they are given permission to mine. The road would go directly to the motorway 25 miles away. Mining noise would be kept to a minimum but blasting would occur every other day at 15:00 hours.

(ii) *Economic and social.* Unemployment in Southshire is slightly above average. Most of the work on local farms and in the seaside resorts is seasonal. In the last few years, the seaside resorts have had fewer visitors.

The people

1 A local Councillor from Shawmouth
2 A teenager from Seaview
3 A young mother from Whitford
4 A farm worker
5 The Secretary of the local environmental group
6 The absentee landlord
7 A representative of the mining company
8 A landlady in Collington

The enquiry

The MP for Southshire has decided to hold a public enquiry to hear the views of the eight people listed above. The enquiry will help to decide whether Zinc UK should be allowed to mine the ore. Suppose that you are one of the people named above. List the advantages and disadvantages that the mining operation might bring to you. Then write out the statement that you would make at the public enquiry.

At present there are otters and deer in Southshire. Should wildlife be disturbed in our quest to use the Earth's resources?

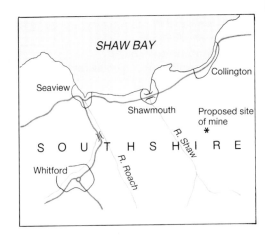

Although mining sometimes damages and scars the countryside, areas can often be restored when finished. This attractive watersports area in the Thames Valley was once a gravel pit. After extracting the gravel for buildings and roads, the pit has been filled with water for sailing and fishing

SECTION E: Study Questions

1 (a) Why was Mendeléev's periodic table so successful and so readily accepted by other scientists?
(b) How do modern periodic tables differ from the one suggested by Mendeléev?

2 This question concerns the following families of elements in the periodic table.
A Group I—the alkali metals
B Group IV—containing carbon and lead
C Group VII—the halogens
D Group O—the noble gases
E The transition elements
Select from A to E, the family containing an element which:
(i) melts at 25 K and boils at 27 K.
(ii) reacts with water forming a solution of pH 12.
(iii) forms a chloride, nitrate and sulphate all of which are pink.
(iv) is a liquid at room temperature and forms a white soluble sodium salt.
(v) forms no compounds.
(vi) is a solid, a poor conductor of electricity and boils above 3000° C.
(vii) is a good conductor of electricity and floats on water.
(viii) can act as a bleach.
(ix) has two crystalline forms at room temperature.
(x) forms an oxide which catalyses the decomposition of hydrogen peroxide very effectively.

3 This question concerns the following outline of the periodic table:

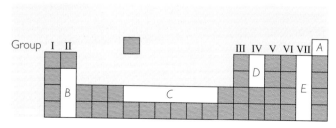

For each of the elements described below, choose the *one* blank area (A to E) in the table above where the element is most likely to be found.
(i) A pale green element which reacts with most metals forming crystalline compounds.
(ii) An element which forms an oxide which dissolves in water to give a solution of pH 9.
(iii) An element which exists as a liquid over a temperature range of less than 10° C.
(iv) An element which reacts with dilute sulphuric acid producing hydrogen but does not react with water.
(v) An element which forms two different solid chlorides, one of which is green and the other yellow.
(vi) An element which is present in sand.
(vii) A colourless, odourless element which is gaseous at room temperature.
(viii) An element with a very high melting point which produces a gaseous acidic oxide.

(ix) A silvery-grey element which reacts vigorously with water producing hydrogen.
(x) An element which reacts with water forming an acidic solution.

4 (a) One important requirement for an anaesthetic is that its boiling point should be between about 40° C and 60° C.
(i) What is the disadvantage of a boiling point below 40° C for an anaesthetic?
(ii) What is the disadvantage of a boiling point above 60° C?
(b) The table below shows the anaesthetic effect, toxicity (i.e., harmful effect on the body) and flammability of methane and its four chlorine-containing derivatives. How do
(i) the anaesthetic effect,
(ii) the toxicity,
(iii) the flammability
depend on the number of chlorine atoms introduced into methane?
(c) Suggest how the chlorine atoms influence the toxicity.

Compound	Anaesthetic effect	Toxicity	Flammability
Methane, CH_4	None	None	Highly flammable
Chloro-methane, CH_3Cl	Weak	Harmful	Highly flammable
Dichloro-methane, CH_2Cl_2	Moderate	Harmful	Non-flammable
Trichloro-methane, $CHCl_3$	Strong	Toxic	Non-flammable
Tetrachloro-methane, CCl_4	Strong	Toxic	Non-flammable

5 The element rubidium (Rb) has similar properties to sodium and potassium and is in the same group in the periodic table.
(a) Write down the formula of:
(i) rubidium oxide
(ii) rubidium hydroxide
(iii) rubidium nitrate.
(b) Describe what happens and write equations for the reactions which occur when:
(i) rubidium is added to water,
(ii) rubidium hydroxide solution is added to hydrochloric acid.
(c) Is rubidium hydroxide an alkali? Explain your answer.

SECTION F
Metals

What properties of metals and alloys made them useful in constructing Concorde?

1 Introducing Metals

Compare the colour of the new coins and the old coins in this photo. Why do coins slowly go darker?

Metals are among the most important and useful materials. Just look around and notice the uses of different metals—cutlery, cars, ornaments, handles and locks for doors, pipes and radiators, girders and bridges.

The reactivity of metals

Shiny aluminium window frames and ladders *quickly become covered* with a thin layer of aluminium oxide.

$$\text{aluminium} + \text{oxygen} \rightarrow \text{aluminium oxide}$$

In the same way, cars and nails made of iron are *slowly* covered with a layer of rust containing iron oxide. Things containing copper, like coins and brass, also react with oxygen in the air to form a thin layer of black copper oxide, but this process is *very slow*.

$$\text{copper} + \text{oxygen} \rightarrow \text{copper oxide}$$

Gold reacts even more slowly than copper. Gold jewellery stays shiny for years. The gold *never reacts* with oxygen in the air. These reactions with oxygen show that aluminium is more reactive than iron, iron is more reactive than copper and copper is more reactive than gold. So, we can now extend the **reactivity series** which we introduced in section B, unit 2 to include six metals (figure 1).

Metals and acids

Various foods, including vinegar, citrus fruits and rhubarb, contain acids. These acids will attack kitchen utensils and cutlery made of certain metals.

Table 1 shows what happens when different metals are added to dilute hydrochloric acid at room temperature (21° C). This acid is more reactive than the acids in foods, but it shows how different metals are attacked.

Wear safety spectacles if you try this experiment and remember that hydrogen is very flammable.

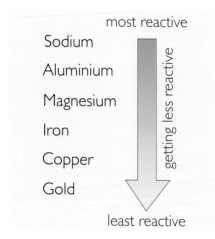

most reactive

Sodium

Aluminium

Magnesium

Iron

Copper

Gold

getting less reactive

least reactive

Figure 1

Which of these foodstuffs contain acids?

One of these spoons has been left in a glass of lemon juice for several days. Citric acid in the juice has attacked metals in the spoon

Notice in table 1 that aluminium does not react at first. This is because its surface is protected by a layer of aluminium oxide. The oxide reacts slowly with the acid. This exposes aluminium, which reacts more vigorously.

The metals used most commonly for pans and cutlery are aluminium, copper, nickel and iron (steel). Copper is the only one which does not react with the acids in food. But copper is so expensive that aluminium and steel are used in most of the saucepans sold today. The thin oxide coating on aluminium helps to protect it from the weak acids in many foods. However, oxalic acid in rhubarb and acetic acid in vinegar do react with aluminium and 'clean' the saucepan during cooking. Iron and nickel, which are used in cutlery, are also attacked by acids in food. Ordinary kitchen cutlery becomes badly marked if it is left in lemon juice (which contains citric acid) or in vinegar.

Foods which contain acids are best stored in unreactive containers made of glass or plastic. Tin cans are also used to store acidic foods like pineapples and grapefruit. Tin cans are made of steel coated on both sides with tin and then lacquered on the inside. The lacquer forms an unreactive layer between the tin and its contents.

Copper is so unreactive that a copper cooking pot will last for many years

Metal used	Reaction with dilute hydrochloric acid	Highest temperature recorded
Aluminium	No reaction at first, but vigorous after a time. Hydrogen is produced rapidly	85° C
Copper	No reaction. No bubbles of hydrogen	21° C
Iron	Slow reaction, bubbles produced slowly from iron	35° C
Lead	Very slow reaction, a few bubbles appear on the lead	23° C
Magnesium	Very vigorous reaction, hydrogen is produced rapidly	95° C
Zinc	Moderate reaction, bubbles of hydrogen are produced steadily	55° C

Table 1: Reactions of metals with dilute hydrochloric acid

Questions

1 Look at table 1.
   (a) Which metal reacts (i) most vigorously, (ii) least vigorously?
   (b) Write the metals in order of decreasing reactivity with dilute hydrochloric acid.
2 Which acids are present in (i) lemon juice, (ii) vinegar, (iii) rhubarb, (iv) sour milk?
3 (a) Why is copper a better metal for saucepans than aluminium?
   (b) Why is copper not used for most pans today?
   (c) Why is aluminium less reactive to acidic foodstuffs than expected?
   (d) How are tin cans protected from acids in their contents?
4 Write word equations for the reactions of
   (a) aluminium with dilute hydrochloric acid,
   (b) magnesium with dilute sulphuric acid,
   (c) zinc with dilute nitric acid.
5 Make a list of six important uses for metals. Say why metals are used in these six ways.

2 The Reactivity Series

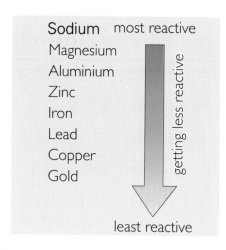

Sodium — most reactive

Magnesium
Aluminium
Zinc
Iron
Lead
Copper
Gold

getting less reactive

least reactive

Figure 1

In the last unit we studied the reactions of metals with dilute hydrochloric acid. All the metals react except copper. The products of the reaction are hydrogen and the chloride of the metal.

metal + hydrochloric acid → metal chloride + hydrogen

Using M as the symbol for the metal, and assuming M forms M^{2+} ions, we can write a general equation as

$$M(s) + 2HCl(aq) \rightarrow MCl_2(aq) + H_2(g)$$

A similar reaction occurs between metals and dilute sulphuric acid. This time the products are hydrogen and a metal sulphate. We can summarize these reactions as

> *metal + acid → metal compound + hydrogen*
> *(above Cu)*

From these reactions of metals with dilute acids and the reactions of metals with air (oxygen), we can draw up an overall **reactivity series**. This is shown in figure 1.

Reactions of metals with water

We can check the order of metals in our reactivity series by investigating the reactions of metals with water. **Safety spectacles should be worn** for these reactions, but you should **not** attempt to react sodium or potassium with water. Table 1 summarizes the reactions of six metals with water.

Potassium reacts violently with water, releasing hydrogen and producing a lilac flame

Metal	Reaction with water
Calcium	Sinks in the water. Steady stream of hydrogen produced. The solution becomes alkaline and cloudy due to the formation of calcium hydroxide
Copper	No reaction
Iron	Rusts very slowly, provided oxygen is present. Brown hydrated iron oxide is produced
Magnesium	Tiny bubbles of hydrogen appear on the surface of the magnesium after a few minutes. The solution slowly becomes alkaline
Potassium	A violent reaction occurs. A globule of molten potassium skates over the water surface, hissing and burning with a lilac flame. Hydrogen and potassium hydroxide form
Sodium	A vigorous reaction occurs. A globule of molten sodium skates about the water surface. Hydrogen and sodium hydroxide form

Table 1: The reactions of six metals with water

Look at the results in table 1.

- Write an order of reactivity for the metals in the table.

- Is the order of reactivity with water similar to that with acids?

- Use the results in table 1 and figure 1 to draw up a reactivity series including the extra metals in table 1.

Summarizing the reactions of metals

Notice that the order of reactivity of metals is the same with air (oxygen), with water and with acids. This is not really surprising because metal atoms react to form metal irons in each case. The higher the metal in the reactivity series, the more easily it forms its ions.

$$M \rightarrow M^{2+} + 2e^-$$

Reaction with air (oxygen). Metals lose electrons to form metal ions. The electrons are taken by oxygen molecules (O_2) forming oxide ions (O^{2-}).

$$\text{metal} + \text{oxygen} \rightarrow \text{metal oxide}$$
$$2M + O_2 \rightarrow 2M^{2+}O^{2-}$$

Reaction with water. Here again, the metals lose electrons to form ions. The electrons are taken by water molecules which form oxide ions (O^{2-}) and hydrogen (H_2).

$$\text{metal} + \text{water} \rightarrow \text{metal oxide} + \text{hydrogen}$$
$$M + H_2O \rightarrow M^{2+}O^{2-} + H_2$$

The oxides of metals high in the reactivity series (e.g. Na_2O and CaO) react further with water to form solutions of their hydroxides.

The reactivity series is a very useful summary of the reactions of metals. Metals at the top of the series want to lose electrons and form ions. Metals at the bottom of the series are just the opposite. Ions of these metals want to gain electrons and form atoms.

What is the disadvantage of using iron (steel) for the boiler in a steam train?

Why are hot water tanks made of copper and cold water tanks made of plastic or steel?

Questions

1 Look at the photographs of the hot water tank and the steam engine. Answer the questions in the captions to the photos.

2 Plan an experiment that you could do to find the position of lithium in the reactivity series. Say what you would do and how you would ensure that your test is fair.

3 (a) *X* is a metal which reacts with dilute hydrochloric acid but not with water. What metal could *X* be?

(b) Metal *P* will remove oxygen from the oxide of metal *Q*, but not from the oxide of metal *R*. Write *P*, *Q* and *R* in their order in the reactivity series (most reactive first).

4 Write word equations and balanced symbolic equations for the reaction of

(a) copper with oxygen,

(b) aluminium with oxygen,

(c) calcium with water,

(d) lead with dilute nitric acid,

(e) magnesium with dilute sulphuric acid.

5 Find out about the manufacture of chromium and titanium. How does the reactivity series help to explain the methods used?

Zinc crystals on the surface of galvanized iron

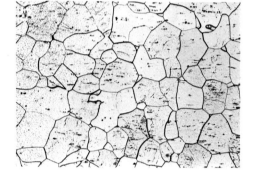

Grain boundaries in antimony

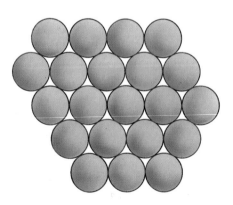

Figure 1

Look carefully at the surface of some galvanized iron (iron coated with zinc). Dustbins, farm gates and cold water tanks are often made of galvanized iron. You will see irregularly-shaped areas separated from each other by clear boundaries. The irregular areas are called **grains** and the boundaries between grains are **grain boundaries**. The grains in tin and zinc are usually easy to see, but the grains in most metals are too small to see with the naked eye. The oxide coating on many metals also makes it difficult to see the grains. But, if the metal surface is clean and smooth, the grains can be seen under a microscope.

X-ray analysis shows that the atoms in metal grains are packed in a regular order, but the grains themselves are irregularly-shaped crystals of the metal pushed tightly together.

Metals usually have a high density. This suggests that the particles are packed close together. The high melting points and boiling points of most metals also indicate that the atoms are held closely together by strong forces of attraction. In fact, X-ray studies show that the atoms of most metals are packed as close together as possible. This arrangement is called **close packing**. Figure 1 shows a few atoms in one layer of a metal crystal. Notice that each atom in the middle of the crystal touches 6 other atoms in the same layer. When a second layer is placed on top of the first layer, atoms in the second layer sink into the dips between atoms in the first layer (figure 2). This means that any one atom in the first layer can touch 6 atoms in its own layer, 3 atoms in the layer above it and 3 atoms in the layer below it. The total number of 'nearest neighbours' to one atom in a close-packed structure is therefore 12. The number of nearest neighbours to an atom or ion in a crystal is called its **coordination number**. Thus, we can say that the coordination number for atoms in a close-packed metal structure is 12.

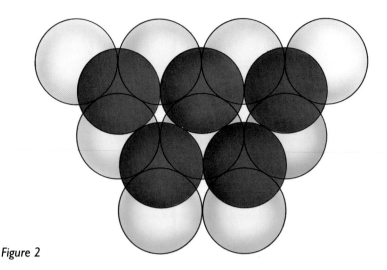

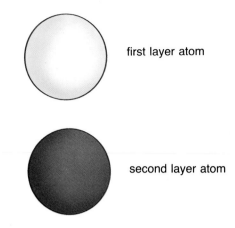

first layer atom

second layer atom

Figure 2

A good model of the close packing in a single layer of metal atoms can be made by blowing a slow stream of gas through a fine jet into a dish containing dilute soap solution (figure 3). The small bubbles look like atoms in a layer of metal.

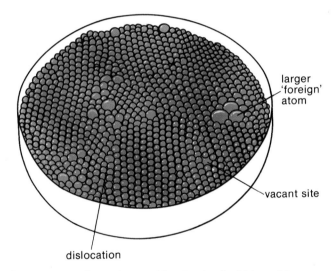

larger 'foreign' atom

vacant site

dislocation

Figure 3
A bubble raft showing crystal grains, and also grain boundaries, dislocations, vacant sites and larger 'foreign' atoms

The large areas of regular packing in the 'bubble raft' represent a single crystal grain. Figure 3 also shows the four main 'flaws' (imperfections) in metal crystals.

1 Grain boundaries. These are the narrow areas of disorder between one crystal grain and another.

2 Dislocations. Occasionally, a row of atoms is displaced or it 'peters out' and the regular packing ceases. When this happens in a metal crystal, the flaw is called a **dislocation**.

3 Vacant sites. These occur where an atom is missing from the crystal structure.

4 'Foreign' atoms. When atoms of another element form part of the crystal structure, they break up the orderly arrangement. This is noticeable in the 'bubble raft' where there are 'foreign' bubbles which are larger or smaller than the rest.

Questions

1 Explain the following:
grain; grain boundary; close packing; coordination number.

2 Name and explain the four main kinds of flaw (imperfection) in metal crystals.

3 (a) Explain why atoms in a close-packed metal structure have a coordination number of 12.
(b) Look closely at the structure of diamond in Unit 3, Section H. What is the co-ordination number of carbon atoms in diamond?
(c) Look closely at the structure of sodium chloride in Unit 8, Section H. What are the co-ordination numbers for Na^+ ions and Cl^- ions in the structure?

4 Properties of Metals

Blacksmiths rely on the malleability of metals to hammer and bend them into useful shapes

The properties of a metal depend on the size of its crystal grains and also on the way its atoms are arranged within the crystal.

> *In general, metals*
>
> - *are shiny solids*
> - *have high densities*
> - *have high melting points*
> - *have high boiling points*
> - *are good conductors of heat*
> - *are good conductors of electricity*
> - *are malleable (can be hammered into different shapes)*
> - *are ductile (can be pulled into wires)*

- **Density.** The close packing of atoms in most metal crystals helps to explain their high densities.

- **Melting and boiling points.** High melting points and high boiling points suggest that there are strong forces holding the atoms together in metal crystals. Chemists think that the outermost electrons of each atom can move about freely in the metal. So, the metal consists of positive ions surrounded by a 'sea' of moving electrons (figure 1). The negative 'sea' of electrons attracts *all* the positive ions and cements everything together. The strong forces of attraction between the moving electrons and the positive ions result in high melting points and high boiling points.

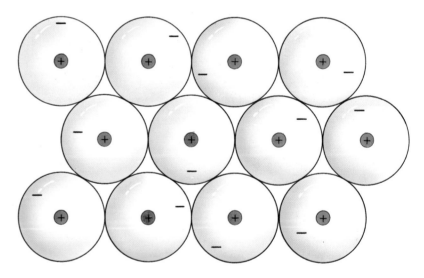

Figure 1
The outermost electrons of each atom move around freely in the metal structure

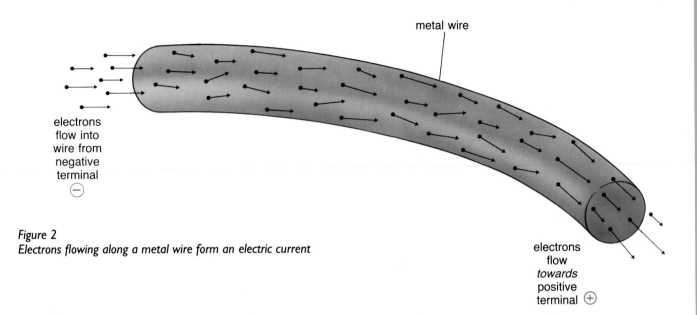

metal wire

electrons
flow into
wire from
negative
terminal
⊖

electrons
flow
towards
positive
terminal ⊕

Figure 2
Electrons flowing along a metal wire form an electric current

• **Conductivity.** When a metal is connected in a circuit, freely moving electrons in the metal move towards the positive terminal. At the same time, electrons can be fed into the other end of the metal from the negative terminal (figure 2). This flow of electrons through the metal forms the electric current.

• **Malleability and ductility.** The bonds between atoms in a metal are strong but they are *not* rigid. When a force is applied to a metal crystal, the layers of atoms can 'slide' over each other. This is known as **slip**. After slipping, the atoms settle into position again and the close-packed structure is restored. Figure 3 shows the positions of atoms before and after slip. It also shows what happens when a metal is hammered into different shapes or drawn into a wire.

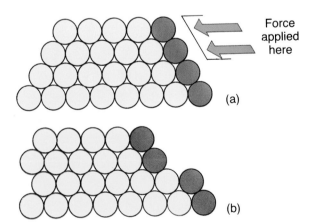

Force
applied
here

(a)

(b)

Figure 3
The positions of atoms in a metal crystal, (a) before and (b) after 'slip' has taken place

---

Questions

1 Explain the words:
slip; malleable; ductile.

2 (a) Blacksmiths often dip red hot steel objects in cold water. Why do they do this?
(Hint: Metals get harder as their crystal grains get smaller.)
(b) Sodium has a much lower density and melting point than most metals. Why is this?

3 Explain why metals (i) have a high density; (ii) have a high melting point; (iii) are good conductors of heat; (iv) are malleable.

4 Answer true or false to parts A to F. Some reasons for classifying magnesium as a metal are
A it burns to form an oxide.
B it reacts with non-metals.
C it reacts only with non-metals.
D it is magnetic.
E it conducts electricity.
F it has a high density.

5 Draw a diagram similar to figure 3 to show how slip occurs when a metal is pulled into a wire.

5 Alloys

Metallurgists have found ways of making metals harder and stronger. They do this by preventing slip in the metal crystals. If slip cannot occur, then the metal is less malleable and less ductile. Metals can be made stronger by reducing the size of crystal grains and by alloying.

Reducing grain size

Slip does not occur across the grain boundaries in a metal. So, metals with small grains are stronger, harder and less malleable than metals with larger grains. Grains can be made smaller by allowing the molten metal to solidify rapidly or by heating the metal and then cooling it quickly (**quenching**).

Alloying

Metals can be made stronger by adding small amounts of another element. Brass is made by mixing copper and zinc. This alloy is much stronger than pure copper or pure zinc. The different-sized atoms break up the regular packing of metal atoms. This prevents slip and makes the metal tougher and less malleable (figure 1). In steel, the lattice of iron atoms is distorted by adding smaller, carbon atoms. These form crystals of iron carbide which are very hard. The regions of iron carbide in the softer iron make steel very strong (figure 2).

As soon as our ancestors had built furnaces that would melt metals, they began to make alloys. The first alloy to be used was probably bronze, a mixture of copper and tin. Bronze swords, ornaments and coins were being made as early as 1500 BC.

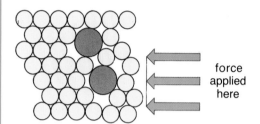

force applied here

Figure 1
Slip cannot occur in the alloy because the atoms of different size cannot slide over each other

Bronze articles like this sword, were made as early as 1500 BC. Bronze is a mixture of copper and tin

Nowadays, we depend on alloys. Most of the metallic materials we use are alloys and not pure metals. Almost all the metal parts in a car are alloys. If the bodywork and components were pure iron, they would rust very rapidly compared to present-day steels. They would also be very malleable and would buckle under strain.

Using alloys

It is possible to make alloys with very specific properties. Some are made for hardness, some are resistant to wear and corrosion. Other alloys have special magnetic or electrical properties. Usually alloys are made by mixing the correct amounts of the constituents in the molten state. For example, solder is made by melting lead, mixing in tin and then casting the alloy into sticks.

Figure 2
Only 0.2% carbon in iron produces steel strong enough for supports and girders in bridges like the Forth Road Bridge

The most important alloys are those based on steel. 'Manganese steel', containing 1% carbon and 13% manganese, can be made very tough by heating it to 1000° C and then 'quenching' (cooling quickly) in water. It is used for parts of rock-breaking machinery and railway lines. Stainless steel is another important alloy. Cutlery contains about 20% chromium and 10% nickel. Chromium stops the steel rusting; nickel makes it harder and less brittle.

2p coins are 97% copper, 2.5% zinc and 0.5% tin. 50p coins are 75% copper and 25% nickel

Steel being quenched

During the last 30 years, aluminium alloys have been used more and more. These include *duralumin* which contains 4% copper. Aluminium alloys are light, strong and corrosion resistant, but they may cost six times as much as steel. They are used for aircraft bodywork, window frames and lightweight tubing. The best-known alloys of copper are brass and bronze. Coins are also copper alloys.

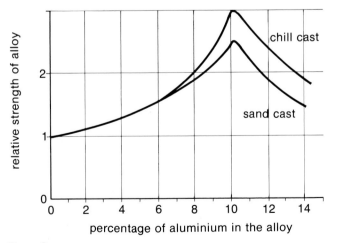

Figure 3
The effect of aluminium on the strength of copper alloys

Questions

1 Name and explain two ways of strengthening metals.
2 Make a table showing the elements present in the following alloys and the main uses of each alloy:
brass; bronze; stainless steel; solder; duralumin.
3 Look at figure 3.
  (a) Which process produces the strongest alloy—chill casting (i.e. rapid cooling) or sand casting (i.e. slow cooling) of the liquid alloy?
  (b) What percentage of aluminium produces the strongest alloy?
  (c) How many times stronger is this alloy than pure copper?
  (d) What percentage of aluminium produces a sand-cast alloy twice as strong as pure copper?
  (e) Why do you think the strength of the alloy increases at first and then decreases as more aluminium is added?
4 Describe an experiment that you could carry out to compare the strengths of two wires made of copper/aluminium alloy.

6 Iron for Industry

A modern blast furnace. The ancient Egyptians were the first people to extract iron from iron ore on a large scale. Today the world production of iron is about 750 million tonnes per year

Iron is the most important metal. Most of it is made into steel and used for machinery, tools, vehicles and large girders in buildings and bridges.

The main raw material for making iron is iron ore (haematite). This ore contains iron(III) oxide. The largest deposits of iron ore in the UK are in Northamptonshire and Lincolnshire, but most of this is of poor quality. The best quality ores are found in Scandinavia, America, Australia, North Africa and Russia.

Iron ore is converted to iron in a special furnace called a **blast furnace** (figure 1). This furnace is a tapered cylindrical tower about 20 metres tall. A mixture of iron ore, coke and limestone is added at the top of the furnace. At the same time, *blasts* of hot air (which give the furnace its name) are blown in through small holes near the bottom. Oxygen in the blast of air reacts with the coke to form carbon monoxide:

$$2C + O_2 \rightarrow 2CO$$
(coke)

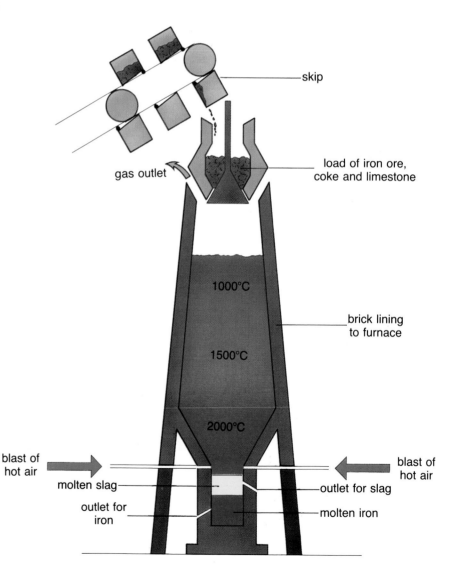

Figure 1
Extracting iron from iron ore in a blast furnace

This reaction is very exothermic and the temperature in the furnace rises to 2000° C.

As the carbon monoxide rises up the furnace, it reduces the iron ore (iron(III) oxide) to iron:

$$Fe_2O_3 + 3CO \rightarrow 2Fe + 3CO_2$$

The high temperatures inside the furnace keep the iron molten. This runs to the bottom of the furnace where it is tapped off from time to time. The molten iron can be used immediately to make steel or it can be poured into moulds to solidify. The large chunks of iron which form are called 'pigs' so the name 'pig-iron' is used for the metal at this stage.

Why is limestone used in the furnace?

Iron ore usually contains impurities like earth and sand (silicon dioxide, SiO_2). Limestone helps to remove these impurities. The limestone decomposes at the high temperatures inside the furnace. It forms calcium oxide and carbon dioxide.

$$CaCO_3 \rightarrow CaO + CO_2$$

The calcium oxide then reacts with sand (SiO_2) and other substances in the impurities to form 'slag' containing calcium silicate:

$$CaO + SiO_2 \rightarrow CaSiO_3$$

The 'slag' falls to the bottom of the furnace and floats on the molten iron. It is tapped off from the furnace at a different level from the molten iron. The 'slag' is used for building materials and cement manufacture.

Blast furnaces work 24 hours a day. Raw materials are continually added at the top of the furnace and hot air is blasted in at the bottom. At the same time, molten slag and molten iron are tapped off from time to time as they collect. This process goes on all the time for about two years. After this time, the furnace has to be closed down so that the lining can be repaired.

Molten iron being tapped from a furnace

Questions

1 (a) Why are furnaces used to make iron called *blast* furnaces?
(b) Why are blast furnaces usually built near coal fields?
(c) What materials are added to a blast furnace?
(d) What materials come out of a blast furnace?

2 (a) The prices of metals are given in the 'commodities' section of the more serious daily papers. Look up the prices for iron (steel), aluminium, copper and zinc. How does the price of iron (steel) compare with other metals?
(b) Suggest 3 reasons why iron is the cheapest of all metals.
(c) Suggest 3 reasons why iron is used in greater quantities than any other metal.

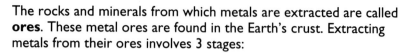

7 Extracting Metals

Mercury can be obtained from cinnabar (impure mercury sulphide) by simply heating the ore

The rocks and minerals from which metals are extracted are called **ores**. These metal ores are found in the Earth's crust. Extracting metals from their ores involves 3 stages:

* mining and concentrating the ore;
* reducing the ore to the metal;
* purifying the metal.

Mining and concentrating the ore

Very often, the ore must be separated from soil and other impurities before it is processed. This may involve crushing the rocks and then separating the ore from impurities by a process of flotation.

Reducing the ore to the metal

The method used for a particular metal depends on:
(i) the position of the metal in the reactivity series and
(ii) the cost of the process.

* **Heating the ore alone**. This is the cheapest way to extract a metal, but it only works with the compounds of *metals at the bottom of the reactivity series*. For example, mercury is extracted from cinnabar (mercury sulphide, HgS) by heating it in air:

$$HgS + O_2 \rightarrow Hg + SO_2$$

* **Reduction with carbon and carbon monoxide.** The *metals in the middle of the reactivity series* such as zinc, iron, tin and lead cannot be obtained simply by heating their ores. They are usually obtained by reducing their oxides with carbon (coke) or carbon monoxide. In the last unit we studied the extraction of iron by reduction of iron(III) oxide using carbon monoxide. In a similar fashion, tin and lead can be extracted from their oxides by reduction with carbon (coke).

$$SnO_2 + 2C \rightarrow Sn + 2CO$$

$$PbO + C \rightarrow Pb + CO$$

Sometimes, these metals exist as sulphide ores. These ores must be converted to oxides before reduction. This is done by heating the sulphides in air.

$$2PbS + 3O_2 \rightarrow 2PbO + 2SO_2$$

* **Electrolysis of molten compounds.** *Metals at the top of the reactivity series*, like sodium, magnesium and aluminium, cannot be obtained by reduction of their oxides with carbon or carbon monoxide. This is because the temperature needed to reduce the oxides is too high. These metals cannot be obtained by electrolysis of their *aqueous* solutions because hydrogen from the water is produced at the cathode instead of the metal. The only way to extract these reactive metals is by electrolysing their *molten* compounds. Potassium, sodium, calcium and magnesium are obtained by electrolysis of their molten chlorides. Aluminium is obtained by electrolysis of molten Al_2O_3 (section D, unit 6).

Tinstone (SnO_2) was once mined in Cornwall and reduced to tin in furnaces near the mines. The tin ore is now virtually used up

Impurity	% in pig-iron	% in mild steel
Carbon	3–5	0.15
Silicon	1–2	0.03
Sulphur	0.05–0.10	0.05
Phosphorus	0.05–1.5	0.05
Manganese	0.5–1.0	0.5–1.0

Table 1: The main impurities in pig-iron and in mild steel

Purifying the metal

Copper straight from the furnace contains 2–3% of impurities. Sheets of this copper are purified by electrolysis with copper sulphate solution (see section D, unit 6). Similarly, pig-iron contains about 8% of impurities (table 1). These impurities make pig-iron very hard and brittle compared to iron and steel. In order to make steel from pig-iron, most of the carbon, silicon, sulphur and phosphorus must be removed. In the basic oxygen process (figure 1), this is done by blowing oxygen onto the hot, molten pig-iron. The oxygen converts carbon and sulphur to carbon dioxide and sulphur dioxide which escape as gases. Silicon and phosphorus are oxidized to solid oxides (P_2O_5 and SiO_2). These combine with lime in the furnace to form slag:

$$3CaO + P_2O_5 \rightarrow Ca_3(PO_4)_2$$
lime, calcium phosphate

$$CaO + SiO_2 \rightarrow CaSiO_3$$
lime, calcium silicate

} slag

Molten iron being poured into a basic oxygen furnace

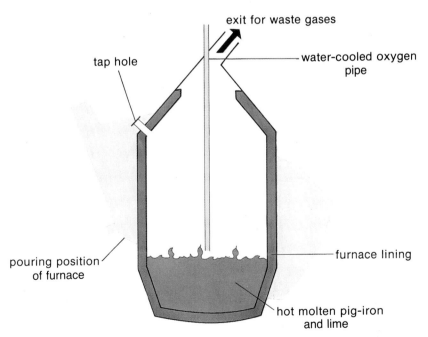

exit for waste gases

tap hole

water-cooled oxygen pipe

pouring position of furnace

furnace lining

hot molten pig-iron and lime

Figure 1
The basic oxygen converter

Questions

1 (a) List the three stages involved in extracting a metal from its ore.
(b) Describe what happens in these three stages when steel is manufactured from iron ore.
2 How does the method used to obtain a metal from its ore depend upon the position of the metal in the reactivity series?
3 Why is iron used in much larger quantities than any other metal?
4 Suggest a method of extracting
(i) silver from silver sulphide (Ag_2S);
(ii) zinc from zinc sulphide (ZnS);
(iii) magnesium from magnesium chloride ($MgCl_2$).

8 Displacing Metals

In the first unit of this section, we studied the reactions of metals with acids. Metals above copper in the reactivity series reacted to form a salt and hydrogen. For example,

$$Zn + H_2SO_4 \rightarrow ZnSO_4 + H_2$$

In this case, zinc has displaced hydrogen from the acid. We can write the equation in terms of ions as

$$Zn(s) + \underbrace{2H^+(aq) + SO_4{}^{2-}(aq)}_{\substack{\text{dilute sulphuric} \\ \text{acid}}} \rightarrow \underbrace{Zn^{2+}(aq) + SO_4{}^{2-}(aq)}_{\substack{\text{zinc sulphate} \\ \text{solution}}} + H_2(g)$$

The sulphate ions take no part in this reaction. We can cancel them from both sides of the equation and write

$$Zn(s) + 2H^+(aq) \rightarrow Zn^{2+}(aq) + H_2(g)$$

Now, if zinc can displace hydrogen ($H_2(g)$) from $H^+(aq)$ ions in acid, then it may be able to displace metals from solutions of their salts. Figure 1 shows what happens when strips of zinc are placed in solutions of various metal ions. Notice that zinc displaces lead from lead nitrate solution, copper from copper sulphate solution and silver from silver nitrate solution. But, zinc does not displace magnesium from magnesium nitrate solution. Table 1 summarizes the results of this experiment and four other experiments in which magnesium, iron, lead and copper are used in place of zinc.

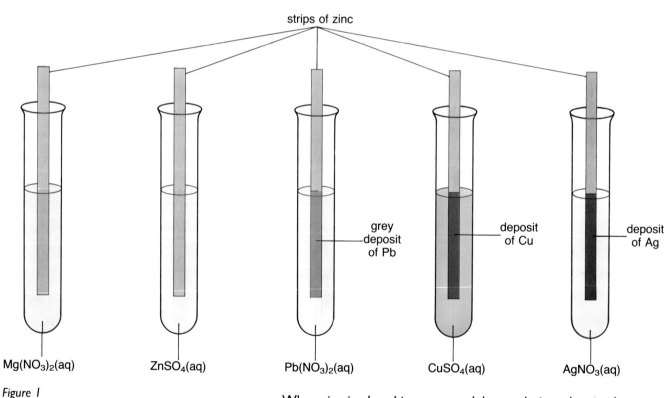

strips of zinc

grey deposit of Pb

deposit of Cu

deposit of Ag

Mg(NO₃)₂(aq)   ZnSO₄(aq)   Pb(NO₃)₂(aq)   CuSO₄(aq)   AgNO₃(aq)

Figure 1

When zinc is placed in copper sulphate solution, the zinc becomes coated with a red brown deposit of copper. At the same time, the blue colour of the solution fades.

Metal used	Results with an aqueous solution of				
	Mg(NO$_3$)$_2$	ZnSO$_4$	Pb(NO$_3$)$_2$	CuSO$_4$	AgNO$_3$
Mg		dark grey deposit of zinc	grey deposit of lead	red-brown deposit of copper	black deposit of silver
Zn	No		grey deposit of lead	red-brown deposit of copper	black deposit of silver
Fe	apparent		grey deposit of lead	red-brown deposit of copper	black deposit of silver
Pb	reaction			red-brown deposit of Cu forms slowly	grey-black deposit of silver
Cu					black deposit of silver

Table 1: The displacement reactions of some metals

Copper ions have been displaced from the solution as copper atoms. These have been deposited on the zinc and zinc ions from the zinc metal have replaced the copper ions in the solution. The equation for the reaction is

$$Zn(s) + Cu^{2+}(aq) + SO_4^{2-}(aq) \rightarrow Zn^{2+}(aq) + SO_4^{2-}(aq) + Cu(s)$$

zinc   copper sulphate   zinc ions in zinc   copper
           solution   sulphate solution

Notice two things in the table:

> **1** The deposits form because the metal added displaces the second metal from a solution of its ions.
>
> **2** The metals and their solutions are written in the order of the reactivity series. *One metal only displaces ions of a metal below it in the reactivity series.* So, zinc can displace lead, copper and silver but not magnesium. Lead can displace copper and silver, but not magnesium and zinc.

These experiments confirm that magnesium is more reactive than the other metals in the table. Magnesium reacts and forms its ions whilst the other metals are displaced from solution. For example,

$$Mg(s) + Pb^{2+}(aq) \rightarrow Mg^{2+}(aq) + Pb(s)$$

Thus, magnesium is higher in the reactivity series than the other metals in table 1. In the same way, zinc is more reactive than lead, copper and silver but less reactive than magnesium. Thus, zinc is above lead, copper and silver in the reactivity series but below magnesium. These results provide further evidence for the order of metals in the reactivity series. They show that metals become less reactive (i.e. less likely to react and form their ions) going down the reactivity series (figure 2).

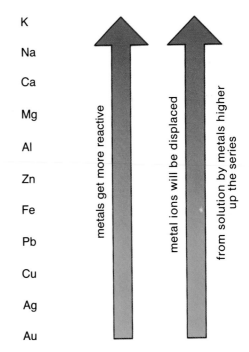

K
Na
Ca
Mg
Al
Zn
Fe
Pb
Cu
Ag
Au

metals get more reactive

metal ions will be displaced

from solution by metals higher up the series

Figure 2

Questions

1 (a) What happens when an iron nail is placed in zinc sulphate solution?
(b) What happens when an iron nail is placed in copper sulphate solution?
(c) Write equations for any reactions in (a) and (b).

2 Iron (steel) is sometimes protected from corrosion by a coating of zinc to form galvanized iron. Use the reactivity series to explain why the zinc protects the iron even when the zinc layer gets scratched and the iron is exposed.

3 Say whether reactions occur in (i) to (v) below. Write equations for any reactions which occur:
(i) *magnesium and aqueous silver nitrate;*
(ii) *silver and aqueous sodium chloride;*
(iii) *copper and aqueous lead nitrate;*
(iv) *aluminium and aqueous iron(II) sulphate;*
(v) *iron and aqueous lead nitrate.*

9 Metal Compounds

Some of the reactions of metal compounds can be linked to the reactivity series like the reactions of metals themselves.

Heating metal compounds

You must wear safety spectacles if you try any of the tests in this unit. The effect of heat on metal compounds can be related to the reactivity series. The equations below summarize what happens when copper hydroxide, copper carbonate and copper nitrate are heated.

$$Cu(OH)_2(s) \rightarrow CuO(s) + H_2O(g)$$
$$CuCO_3(s) \rightarrow CuO(s) + CO_2(g)$$
$$Cu(NO_3)_2(s) \rightarrow CuO(s) + 2NO_2(g) + \frac{1}{2}O_2(g)$$

Notice how similar the equations are. In each case, the metal compound decomposes to a solid metal oxide and a gaseous non-metal oxide:

$$\text{metal compound} \xrightarrow{heat} \text{metal oxide} + \text{non-metal oxide}$$

Other common metal compounds—sulphates, sulphides and chlorides, are more stable than hydroxides, carbonates and nitrates, and they do not decompose so readily on heating.

The action of heat on metal hydroxides, carbonates and nitrates is summarized in table 1.

Metal	Action of heat on hydroxide	Action of heat on carbonate	Action of heat on nitrate
K Na	stable	stable	decompose to nitrite + O_2 $NaNO_3 \rightarrow$ $NaNO_2 + \frac{1}{2}O_2$
Ca Mg Al Zn Fe Pb Cu	decompose to oxide + H_2O e.g. $Ca(OH)_2 \rightarrow$ $CaO + H_2O$	decompose to oxide + CO_2 e.g. $CaCO_3 \rightarrow$ $CaO + CO_2$	decompose to oxide + $NO_2 + O_2$ e.g. $Ca(NO_3)_2 \rightarrow$ $CaO + 2NO_2 + \frac{1}{2}O_2$
Ag Au	hydroxides are too unstable to exist	carbonates are too unstable exist	decompose to metal + $NO_2 + O_2$ $AgNO_3 \rightarrow$ $Ag + NO_2 + \frac{1}{2}O_2$

Table 1: The action of heat on metal compounds

Notice that in Table 1 the metal compounds fall into three groups.

1 The most reactive metals, potassium and sodium, have the most stable compounds. Only the nitrates of these metals decompose on heating with a Bunsen (which reaches about 900° C).

2 Metals in the middle of the reactivity series from calcium to copper form less-stable compounds. Their hydroxides, carbonates and nitrates decompose to give the metal oxide on heating.

Lime (calcium oxide) has been produced since the 18th century by heating limestone. This photo shows a lime burning kiln in the Carpathian Mountains, Hungary

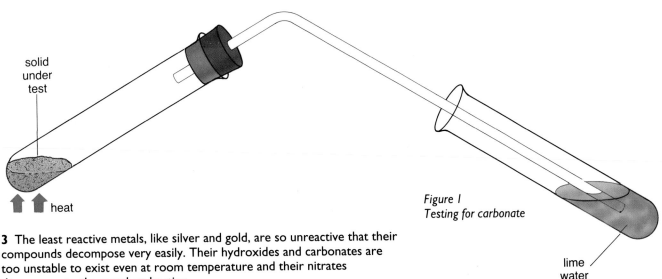

Figure 1
Testing for carbonate

lime
water

solid
under
test

heat

3 The least reactive metals, like silver and gold, are so unreactive that their compounds decompose very easily. Their hydroxides and carbonates are too unstable to exist even at room temperature and their nitrates decompose to the metal on heating.

> *In general, the lower a metal is in the reactivity series, the easier it is to decompose its compounds.*

Identifying anions

The action of heat on metal compounds can be used to identify most carbonates (CO_3^{2-}) and nitrates (NO_3^-).

- **Test for carbonate.** Heat the substance and pass any gases produced on heating through lime water (figure 1). If the solid is a carbonate, carbon dioxide may be produced. This turns lime water milky.

  In order to make sure that the substance is a carbonate, add dilute nitric acid to the solid. Carbon dioxide should be produced:

$$CO_3^{2-}(s) + 2H^+(aq) \rightarrow CO_2 + H_2O$$

- **Test for nitrate.** Most nitrates give brown fumes of nitrogen dioxide (NO_2) on heating. This is good evidence for a nitrate. In order to check that the substance is a nitrate, add fresh iron(II) sulphate solution to an aqueous solution of it. Then, **carefully (wearing eye protection)** pour concentrated sulphuric acid down the inside of the test tube (figure 2). A brown ring should form where the acid and the aqueous layer meet.

- **Test for chloride.** Add dilute nitric acid to a solution of the substance. Then add silver nitrate solution. If the substance contains chloride (Cl^-), a white precipitate of silver chloride forms, which turns purple grey in sunlight:

$$Ag^+(aq) + Cl^-(aq) \rightarrow AgCl(s)$$

- **Test for sulphate.** Add dilute nitric acid to a solution of the substance. Then add barium nitrate solution. If the substance contains sulphate (SO_4^{2-}), a thick white precipitate of barium sulphate ($BaSO_4$) forms:

$$Ba^{2+}(aq) + SO_4^{2-}(aq) \rightarrow BaSO_4(s)$$

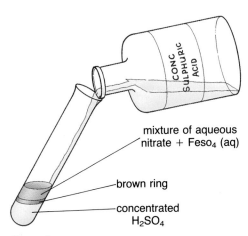

CONC SULPHURIC ACID

mixture of aqueous
nitrate + $FeSO_4$ (aq)

brown ring

concentrated
H_2SO_4

Figure 2
Testing for nitrate

Questions

1 How is the decomposition of metal carbonates related to the reactivity series?

2 Write equations for the action of heat on the following compounds. If decomposition does not occur, say so:
$Zn(OH)_2$; KOH; Ag_2O; Rb_2CO_3; $PbCO_3$; $Mg(NO_3)_2$.

3 Find out about the manufacture of lime from limestone. What are the uses of (i) limestone, (ii) lime?

4 Two fertilizers are labelled 'sulphate of potash' and 'nitrate of potash'. How would you check which was which?

10 Identifying Cations

Cation present in substance	Flame colour
K^+	lilac
Na^+	yellow
Ca^{2+}	red
Cu^{2+}	blue/green

Table 1: The flame colours from some cation

We can identify different substances using tests for the atoms and ions in them. The best way to identify a substance is to find a property or a reaction which is shown only by that substance and is easy to see.

Two of the best tests for cations use flame tests and sodium hydroxide solution. **You must wear eye protection** if you try any of the tests in this unit.

Flame tests

When substances are heated strongly, the electrons in them absorb extra energy. We say that the electrons are **excited**. Very soon, the excited electrons give out this excess energy as light and they become stable again. When this happens, different cations are found to emit different colours of light (table 1).

See if you can get a good flame test for a potassium or a copper compound using the following method.

Dip a nichrome wire in concentrated hydrochloric acid and then heat it in a roaring Bunsen *until it gives no colour to the flame*. The wire is now clean. Dip it in concentrated hydrochloric acid again, and then in the substance to be tested. Heat the wire in the Bunsen and note the flame colour.

Tests using sodium hydroxide solution

The hydroxides of all metals (except those in Group 1) are insoluble. So, these insoluble hydroxides form as solids (precipitates) when aqueous sodium hydroxide, containing hydroxide ions (OH^-), is added to a solution of metal ions. For example:

$$Cu^{2+}(aq) + 2OH^-(aq) \rightarrow Cu(OH)_2(s)$$
$$\text{blue precipitate}$$

Cation in solution	3 drops of NaOH(aq) added to 3 cm^3 of solution of cation	10 cm^3 of NaOH(aq) added to 3 cm^3 of solution of cation
K^+	No precipitate	No precipitate
Na^+	No precipitate	No precipitate
Ca^{2+}	A white precipitate of $Ca(OH)_2$ forms $Ca^{2+} + 2OH^- \rightarrow Ca(OH)_2$	White precipitate remains
Mg^{2+}	A white precipitate of $Mg(OH)_2$ forms $Mg^{2+} + 2OH^- \rightarrow Mg(OH)_2$	White precipitate remains
Al^{3+}	A white precipitate of $Al(OH)_3$ forms $Al^{3+} + 3OH^- \rightarrow Al(OH)_3$	White precipitate dissolves to give a colourless solution
Zn^{2+}	A white precipitate of $Zn(OH)_2$ forms $Zn^{2+} + 2OH^- \rightarrow Zn(OH)_2$	White precipitate dissolves to give a colourless solution
Fe^{2+}	A green precipitate of $Fe(OH)_2$ forms $Fe^{2+} + 2OH^- \rightarrow Fe(OH)_2$	Green precipitate remains
Fe^{3+}	A brown precipitate of $Fe(OH)_3$ forms $Fe^{3+} + 3OH^- \rightarrow Fe(OH)_3$	Brown precipitate remains
Cu^{2+}	A blue precipitate of $Cu(OH)_2$ forms $Cu^{2+} + 2OH^- \rightarrow Cu(OH)_2$	Blue precipitate remains

Table 2: Testing for cations with sodium hydroxide solution

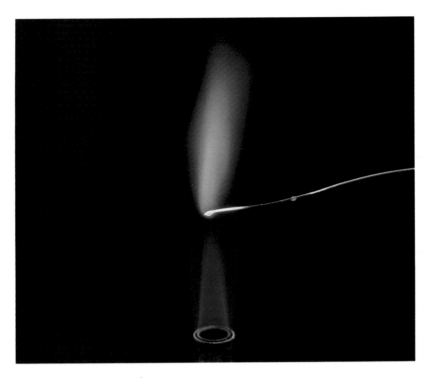

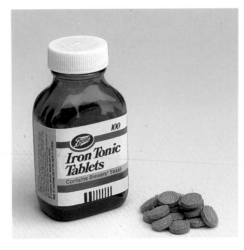

What tests would you carry out to see whether iron tablets contain Fe^{2+} or Fe^{3+} ions?

Carrying out a flame test for potassium

Table 2 shows what happens when
(i) a little sodium hydroxide solution, NaOH (aq), and then
(ii) excess sodium hydroxide solution is added to solutions of some common cations.
Look carefully at table 2.

1 Which cations give no precipitate with sodium hydroxide solution? How can you tell the difference between these cations using flame tests?

2 Which cations give a white precipitate which does *not* dissolve in excess sodium hydroxide solution? How can you tell the difference between these cations using flame tests?

3 Which cations give a coloured (non-white) precipitate with sodium hydroxide solution? How can you tell the difference between these cations from the colour of their hydroxides?

4 Which cations give a white precipitate which dissolves in excess sodium hydroxide solution? These metal ions can be identified by heating their hydroxides to dryness:

$$2Al(OH)_3 \overset{heat}{\rightarrow} Al_2O_3 + 3H_2O$$
white

$$Zn(OH)_2 \overset{heat}{\rightarrow} ZnO + H_2O$$
yellow
when hot

When aluminium hydroxide is heated, it produces a white oxide. When zinc hydroxide is heated, it produces an oxide which is yellow when hot, but white on cooling.

Questions

1 (a) How would you carry out a flame test on a sample of chalk?
(b) What colour should the flame test give?
(c) Why must eye protection be worn during flame testing?
(d) What causes flame colours from certain cations?

2 (a) A garden pesticide is thought to contain copper sulphate. Describe two tests that you would do to find out whether it contains Cu^{2+} ions.
(b) How would you show that the pesticide contains sulphate?

3 A metal has become coated with a layer of oxide. How would you check whether the metal is aluminium or zinc?

Section F: Activities

1 The importance of iron and steel

Iron is the most important metal for industry and society. There are four reasons for this.

- **It is strong.** Although pure iron is weak, it can be alloyed with carbon to make steel, which is very strong. Steel has a high tensile strength. This means that it does not crack, break, stretch or bend easily. Large forces can deform steel, but it returns to its original shape when the force is removed.

- **It is easy to work.** Iron is very ductile and very malleable when it is hot. This makes it very useful for manufacturing different articles.

- **It forms alloys.** Iron can form a large range of alloys with a wide variety of different properties.

- **It is cheap.** Iron is the cheapest of all metals.

Because of these four factors, iron is used in greater quantity than any other metal.

1 List the factors which make iron more important than any other metal.
2 What do the following mean?
(i) tensile strength
(ii) ductile
(iii) malleable
3 The prices of metals are given in the 'commodities' section of some daily papers and also on teletext. Look up the prices for iron (steel), aluminium, copper and zinc. How does the price of iron (steel) compare with the other metals?
4 Suggest three reasons why iron is the cheapest of all metals.
5 List five important uses of iron (steel).

The Eiffel Tower is made of steel. Why is steel used? What other important structural materials do we have for large buildings and bridges?

2 Panning for gold

Look closely at the drawing of prospectors panning for gold in California.

1 Why does the gravel in the bed of the stream contain gold?
2 Why is the gold found as the element and not as compounds of gold?
3 What method did the prospectors use to separate gold from other materials in the gravel?
4 Why did the method in question 3 separate the gold?
5 What problems do you think the gold prospectors faced from day to day?

3 | Mining and producing copper

Look carefully at the information below.

Copper

What is the main mineral ore? Copper pyrites (chalcopyrites) a mixture of CuS and FeS forming $CuFeS_2$.

How is it found? Underground, mixed with other copper ores such as chalcocite (Cu_2S). About 0.5% of the ore is copper.

Where is it mined? One third of all copper ores mined in the West comes from Chile and the USA. The other major mining countries are Canada, Zambia and Zaire.

How is it mined? By open pit mining or underground mining.

What is the world production of copper? See table 1. About one seventh of the copper produced is recycled from scrap. This is called **secondary copper**. Copper produced from its ores is called **primary copper**.

Country	Production in tonnes
USA	1 500 000
Japan	935 000
Chile	879 000
Zambia	522 000
Canada	505 000
Belgium	400 000
West Germany	379 000
UK (secondary copper only)	140 000

Table 1: The major producers of copper shown in order. The figures are from 1984

Mining copper ore in Arizona. The different colours are due to malachite (green copper(II) carbonate), azurite (blue hydrated copper(II) carbonate), and cuprite (red copper(I) oxide)

1 What are the chemical names for
(a) CuS and FeS,
(b) Cu_2S?

2 What percentage of the ore is copper?

3 What mass of the ore is needed to produce a 1 kg piece of copper?

4 Only a small percentage of the ore is copper. Suggest three problems that this might cause.

5 The copper ore mined in Chile in 1984 was sufficient to produce nearly 1 300 000 tonnes of copper.
(a) How much copper did Chile actually produce in 1984?
(b) How do you account for the difference between the actual copper produced and the possible production of 1 300 000 tonnes?

6 What is the difference between primary copper and secondary copper?

7 Look at the countries where copper is mined. Which country is surprisingly absent from the major producers of the metal?

8 Make a bar chart showing the total copper production in 1984 for the seven most important producers.

Section F: Study Questions

I (a) From the reactivity series choose a metal that
(i) occurs uncombined in the Earth's crust
(ii) reacts vigorously with cold water
(iii) is used in flares.
(b) Tin (Sn) is slightly more reactive than lead. Predict what you would see when tin foil is placed in dilute hydrochloric acid.
(c) A gas may be made by passing steam over heated zinc using the apparatus shown below:

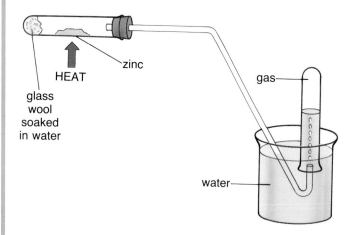

(i) Name the gas collected.
(ii) Write a word equation and a symbolic equation for the reaction.
(iii) Why should you wear safety goggles when carrying out this experiment?
(iv) Which substance has acted as a reducing agent in this reaction? Explain your choice.
(d) Chromium (Cr) may be prepared by heating chromium(III) oxide with aluminium powder.
(i) What does this tell you about the relative positions of chromium and aluminium in the reactivity series?
(ii) Write a word equation and a symbolic equation for the reaction.

2 Consider the following five alloys, A, B, C, D and E.

Alloy	Elements in the alloy	Properties
A	lead and tin	melts at 203° C
B	bismuth, cadmium, lead and tin	melts at 70° C
C	carbon, iron and tungsten	unaffected at high temperatures
D	copper and zinc	golden colour, does not tarnish
E	aluminium and lithium	low density, high strength

Which alloy would you use for
(a) joining electrical wires?
(b) making jewellery?
(c) 'plugging' an automatic fire sprinkler?
(d) making aircraft bodywork?
(e) making a drill for bricks and stone?

3 Look at the table below. It shows metals and alloys which are used in different parts of a car. For each of the parts listed, say why that particular metal or alloy is used.

Car part/component	Metal/alloy used
Bodywork and petrol tank	mild steel (iron with 0.15% carbon)
Pistons	aluminium alloy
Door handles, surrounds to headlamps	chromium-plated steel
Contact breakers	platinum
Lamp filaments	tungsten
Radiator	copper

4 The labels have become unreadable on three bottles that were known to contain
ammonium chloride
calcium carbonate
iron(II) sulphate.
 What *chemical* tests would you carry out to enable you to re-label the bottles correctly? Your answers must include at least *one positive*, chemical test for each substance that would not be given by either of the other two.

5 (a) Name *one* natural form of calcium carbonate.
(b) Describe tests you might use to prove that a piece of rock contained
(i) carbonate ions
(ii) calcium ions.
(c) Strontium (Sr) is in the same group of the periodic table as calcium. Write the formula of strontium nitrate.

Acids, Bases and Salts

Oranges and other fruits contain citric acid. This gives them a sharp taste

1 Introducing Acids

The sharp taste of lemon juice and vinegar is caused by acids. Some acids can be identified by their sharp taste, but it would be dangerous to rely on taste for all acids. Think of the dangers of tasting sulphuric acid or nitric acid.

A more sensible test for acids is to use indicators like litmus and universal indicator. Indicators change colour depending on how acidic or how alkaline a solution is.

● Acidic solutions turn litmus red and give an orange or red colour with universal indicator (figure 1),

● Alkaline solutions turn litmus blue and give a green, blue or violet colour with universal indicator.

It would be awkward to use the colour of an indicator to describe how acidic something is. So chemists use a scale of numbers called the **pH scale**. The pH of a solution gives a measure of its acidity or alkalinity. On this scale:

● Acidic solutions have a pH below 7.
● Alkaline solutions have a pH above 7.
● Neutral solutions have a pH of 7.

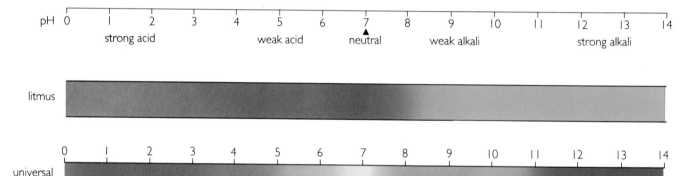

Figure 1
The colours of litmus and universal indicator with solutions of different pH

What protective clothing is this technician wearing to protect herself from the acid she is pipetting?

Making acids

● **From non-metal oxides and water.** When non-metal oxides are added to water, acids are produced. For example:

$$\text{carbon dioxide} + \text{water} \rightarrow \text{carbonic acid}$$
$$CO_2 + H_2O \rightarrow H_2CO_3$$

$$\text{sulphur dioxide} + \text{water} \rightarrow \text{sulphurous acid}$$
$$SO_2 + H_2O \rightarrow H_2SO_3$$

Sulphuric acid is manufactured via sulphur dioxide. The raw materials are sulphur, air (oxygen) and water. Most of the sulphur is shipped to the UK from North America.

There are three stages in the manufacture of sulphuric acid.

1 First, sulphur dioxide is made by burning sulphur in air (oxygen).
$$S + O_2 \rightarrow SO_2$$

A photograph of a sulphuric acid plant. The hot gas filter removes unburnt sulphur dust from sulphur dioxide produced in the first stage of the process. What do you think happens in the converter? The heat exchanger uses the heat produced in the contact process to warm up the reactant gases. Why is this sensible?

2 The sulphur dioxide is then converted to sulphur trioxide. This is called **the contact process**. The sulphur dioxide is mixed with air (oxygen) and passed over a catalyst (vanadium(V) oxide) at 450° C.

sulphur dioxide + oxygen → sulphur trioxide
$$2SO_2 + O_2 \rightarrow 2SO_3$$

3 Finally, the sulphur trioxide is converted to sulphuric acid. Sulphur trioxide does not dissolve very easily in water. So, sulphur trioxide is dissolved in 98% sulphuric acid and then water is added.

$$SO_3 + H_2O \rightarrow H_2SO_4$$

- **By combining hydrogen with reactive elements.** The most important acid to be manufactured in this way is hydrochloric acid. Hydrogen chloride is first made by burning hydrogen in chlorine.

$$H_2 + Cl_2 \rightarrow 2HCl$$

The hydrogen chloride is then passed up a tower packed with stones (or some other unreactive substance) over which water trickles. The gas dissolves in the water to form hydrochloric acid.

Sulphur being mined in an open pit in North America. Most sulphur for the manufacture of sulphuric acid in the U.K. is imported from North America

Questions

1 (a) How would you check the pH of toothpaste?
(b) Why is the pH of toothpaste important?
2 Why are acids important?
3 This question is about the manufacture of sulphuric acid.
  (a) Write equations for the three stages in the manufacture of sulphuric acid.
  (b) What is the contact process?
  (c) Suggest two reasons why a catalyst is used in the contact process.
  (d) 450° C is an optimum temperature for the contact process. What do you think the term 'optimum temperature' means?
  (e) Sulphur trioxide from the contact process is not reacted directly with water to make sulphuric acid. Explain why.
  (f) Most of the sulphuric acid plants in the UK are situated near large ports. Suggest reasons for this.

2 Acids in Everyday Life

Acids are important in everyday life and in industry. They are used to make foods, clothes and medicines.

Acids in our foods

Many foods and drinks contain acids. Citrus fruits—oranges, lemons, pineapples, grapefruit—contain citric acid. Tomato sauce, brown sauce and mint sauce get their sharp taste from vinegar which contains acetic acid.

One of the simplest and cheapest drinks is soda water. This is made by dissolving carbon dioxide in water under pressure. Some of the carbon dioxide reacts with the water to form carbonic acid. When soda water is poured from the bottle, it fizzes because the pressure is lower, which makes carbon dioxide form as bubbles in the drink. Other fizzy drinks, like Coke and lemonade, also contain carbonic acid.

Acids in our bodies

Our stomach walls produce hydrochloric acid. This gives a pH of about 2 in the stomach. These acidic conditions help us to **digest** (break down) foods, particularly proteins. Proteins are broken down into smaller molecules, called peptides and amino acids. These small molecules can pass through the lining of our intestines into the bloodstream. The blood carries these small molecules round the body to where they are needed.

The pH of our blood is 7.4 (slightly alkaline). Most reactions in our bodies can only take place within a narrow range of pH. A change in the pH in your stomach might give you indigestion, but a pH change of only 0.5 in the blood would probably kill you. In order to prevent changes in pH, our bodies contain substances that counteract the effects of acids and alkalis.

Carbonic acid in drinks like Coca Cola, lemonade and soda water makes them 'fizz'

Acid rain, caused by fumes from industry has had drastic effects on the natural environment of Czechoslovakia. Trees have been stunted or killed and the decrease in the pH of lakes has wiped out whole food chains

Acidic pollutants in the Sea of Galilee have killed these fish

Acids in the air—acid rain

When fuels burn, sulphur dioxide and carbon dioxide are produced. Because of this, city air may contain five times as much sulphur dioxide as fresh air. Sulphur dioxide and carbon dioxide react with water vapour and rain in the air to form sulphurous acid, H_2SO_3, and carbonic acid, H_2CO_3.

$$SO_2 + H_2O \quad \rightarrow \quad H_2SO_3$$
$$\text{(sulphurous acid)}$$

$$CO_2 + H_2O \quad \rightarrow \quad H_2CO_3$$
$$\text{(carbonic acid)}$$

These acids make the rain acidic, which is why we use the term 'acid rain'. (See section B, unit 4.)

Many foods especially sauces and pickles, contain acetic acid from vinegar

Most heather grows best in acid soil

These beech trees grow best in slightly alkaline soil

Acids in the soil

The pH of soil can vary from about 4 to 8, but most soils have a pH of between 6.5 and 7.5. In chalk or limestone areas, the soil is usually alkaline, but in moorland, sandstone and forested areas it is generally acidic. Peat bogs and clay soils are also normally acidic. For general gardening and farming purposes, the best results are obtained from a neutral or slightly acidic soil of pH 6.5 to 7.0. Only a few plants, including rhododendrons and azaleas, can grow well in soils which are more acidic than pH 6.5. In areas where the soil is too acidic, it can be improved by treatment with powdered limestone (calcium carbonate) or slaked lime (calcium hydroxide). These substances react with acids in the soil and raise the pH to the right value.

Questions

1 (a) Describe how you would check the pH of soil samples.
  (b) Why do you think that forested and peaty areas have acid soils?
  (c) Why is it important for gardeners to know the pH of the soil?
2 Make a list of foodstuffs which are acidic. Try to find out what acids they contain.
3 (a) Why is soda water both fizzy and acidic?
  (b) Would you expect mineral waters to be (i) fizzy, (ii) acidic? Explain your answers.
  (c) Design an experiment to compare the fizziness and pH of different soda waters and mineral waters.
4 Find out more about acid rain. Here are some questions to consider.
  (a) When did acid rain first become a problem?
  (b) Why did this problem not arise earlier?
  (c) What are the major factors causing acid rain?
  (d) What steps might be taken to reduce acid rain?
Prepare a short report for your class.

3 Neutralization

What substances in 'Rennies' help to neutralize acids in the stomach?

Adverts for indigestion cures usually talk of 'acid stomach' and 'acid indigestion'. Medicines which ease stomach ache (such as Milk of Magnesia, Setlers and Rennies) are called antacids (anti-acids) because they neutralize excess acid in the stomach.

- Substances which neutralize acids are called **bases**. Bases which are soluble in water are called **alkalis**. The bases in indigestion tablets include magnesium hydroxide and calcium carbonate.
- The reactions between acids and bases are called **neutralizations**.

As we saw in the last section, farmers use slaked lime to neutralize acid soils. The main base in slaked lime is calcium hydroxide.

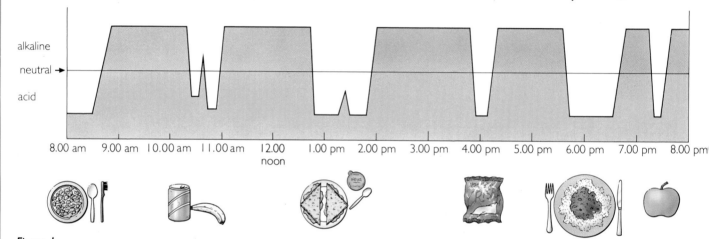

Figure 1
This shows how the pH in your mouth changes during the day. Notice how the pH becomes acidic during and just after meals. This is because sugar and other chemicals in food are being broken down into acids. These acids can cause tooth decay

'Nitram' fertilizer is ammonium nitrate. It is manufactured by a neutralization reaction

Neutralization is also important in dental care. We use toothpaste to neutralize the acids produced when food breaks down in the mouth (figure 1). Neutralization is used to treat splashes of acid or alkali on people's skin or clothes. The stings from many plants and animals contain acids or bases so they can also be treated by neutralization. For example, nettle stings and wasp strings contain bases like **histamine**. They are treated with acidic substances called **anti-histamines**. Unlike wasp stings, those caused by bees are acidic and should be treated with bases.

Neutralization is used to remove carbon dioxide from the air in air-conditioned buildings. Carbon dioxide is an acidic oxide. So the air is passed over soda lime, a mixture of two bases (sodium hydroxide and calcium hydroxide). These react with the carbon dioxide and neutralize the air.

One of the most important applications of neutralization is in making fertilizers like ammonium sulphate ((NH_4)$_2SO_4$) and ammonium nitrate (NH_4NO_3). Ammonium sulphate is manufactured by neutralizing sulphuric acid (H_2SO_4) with ammonia (NH_3). Ammonium nitrate is manufactured by neutralizing nitric acid (HNO_3) with ammonia (NH_3).

Investigating neutralization

When acids are neutralized by bases, the pH of the solution changes. This pH change can be followed using an indicator.

Wear eye protection if you try the following experiments. Put 10 cm^3 of dilute hydrochloric acid ($HCl(aq)$) in a conical flask and add 4 or 5 drops of universal indicator. The colour is red showing a pH of 2 or 3. Now add dilute sodium hydroxide ($NaOH(aq)$) to the mixture, 2 cm^3 at a time, until the solution becomes alkaline and the mixture is a blue/purple colour.

Notice that the base (sodium hydroxide) has neutralized the hydrochloric acid, and by the end is in excess.

We can write an equation for the reaction as:

$$HCl(aq) + NaOH(aq) \rightarrow NaCl(aq) + H_2O(l)$$

Substances like sodium chloride which form when acids react with bases are called **salts**. Sodium chloride is usually known as common salt. Other important salts include copper sulphate ($CuSO_4$), potassium chloride (KCl) and ammonium nitrate (NH_4NO_3).

When hydrochloric acid is electrolysed, it produces hydrogen at the cathode and chlorine at the anode. This shows that it contains H^+ ions and Cl^- ions. So, we can write $H^+(aq) + Cl^-(aq)$ in place of $HCl(aq)$ in equations.

Similarly, sodium hydroxide ($NaOH(aq)$) and sodium chloride ($NaCl(aq)$) also consist of ions. They can be written as

$$Na^+(aq) + OH^-(aq) \text{ and } Na^+(aq) + Cl^-(aq).$$

Water does not consist of ions so it must be written as H_2O.

We can now rewrite the last equation in terms of ions as:

$$H^+(aq) + Cl^-(aq) + Na^+(aq) + OH^-(aq) \rightarrow Na^+(aq) + Cl^-(aq) + H_2O(l)$$

Notice that $Na^+(aq)$ and $Cl^-(aq)$ appear on both sides of this equation. These two ions do not really react. They are just the same after the reaction as they were before. They are called **spectator ions**. If we cancel $Na^+(aq)$ and $Cl^-(aq)$ on both sides of the equation, we get:

$$H^+(aq) + OH^-(aq) \rightarrow H_2O(l)$$

This final equation shows clearly what happens. H^+ ions in the acid have reacted with OH^- ions in the base to form water. Na^+ and Cl^- ions have just 'stood by and watched' the reaction like spectators at a sports fixture.

Naming salts

Look at table 1. This shows how the name of a salt comes from its parent base and acid. So magnesium oxide and nitric acid would produce magnesium nitrate. Notice also in table 1 that

● acids with names ending in '-ic' form salts with names ending in '-ate'. (Hydrochloric acid is an exception—its salts are called chlorides.)

● acids with names ending in '-ous' form salts with names ending in '-ite'.

Base	Name of salt
Magnesium oxide	Magnesium—
Potassium hydroxide	Potassium—
Zinc oxide	Zinc—

Acid	Name of salt
Sulphuric	—sulphate
Sulphurous	—sulphite
Nitric	—nitrate
Nitrous	—nitrite
Carbonic	—carbonate
Hydrochloric	—chloride

Table 1: The names of salts

Questions

1 Design an experiment to compare the effectiveness of indigestion tablets. (Which tablets are best at neutralizing acid? Which tablets are the best buy?)

2 (a) What causes indigestion?
(b) Why can calcium carbonate be classified as both a salt and a base?
(c) Why is neutralization important?

3 (a) Why is ammonium nitrate (NH_4NO_3) used as a fertilizer?
(b) What substances are used to manufacture ammonium nitrate?
(c) Write an equation for the reaction to make ammonium nitrate.

4 Complete the following.
(a) Magnesium oxide + sulphuric acid → __?__ + __?__
(b) Potassium hydroxide + __?__ → __?__ nitrate + water
(c) __?__ + hydrochloric acid → zinc __?__ + __?__

4 Properties of Acids

Car batteries contain sulphuric acid. H^+ and SO_4^- ions in the acid react with lead plates in the battery to produce electricity. This electricity starts the engine and powers things like the radio, lights and windscreen wipers

We know a lot about acids already.

1 Acids are soluble in water. Their solutions have a pH below 7.

2 Acids give characteristic colours with indicators. They give a red colour with litmus and an orange or red colour with universal indicator.

3 Acids react with metals above copper in the reactivity series to form a salt and hydrogen.

| metal + acid → salt + hydrogen |

e.g. $Mg + H_2SO_4 \rightarrow MgSO_4 + H_2$

Nearly all salts contain a metal and at least one non-metal. They are ionic compounds with a cation (e.g. Cu^{2+}, Mg^{2+}) and an anion (Cl^-, SO_4^{2-}). Sodium chloride (Na^+Cl^-) is known as common salt.

4 Acids react with bases (such as metal oxides and hydroxides) to form a salt and water. These reactions are called **neutralizations**.

| base + acid → salt + water |

For example, when black copper oxide is added to warm dilute sulphuric acid, the black solid disappears and a blue solution of copper sulphate forms.

$$CuO + H_2SO_4 \rightarrow CuSO_4 + H_2O$$

5 Acids react with carbonates to give a salt plus carbon dioxide and water.

| carbonate + acid → salt + CO_2 + H_2O |

For example, when sodium carbonate is added to hydrochloric acid, sodium chloride forms and bubbles of carbon dioxide are produced.

$$Na_2CO_3 + 2HCl \rightarrow 2NaCl + CO_2 + H_2O$$

6 Acid solutions conduct electricity and are decomposed by it. This shows that solutions of acids consist of ions (table 1). All acids produce hydrogen at the cathode during electrolysis. This shows that all acids contain H^+ ions. Because of this

| acids are defined as substances which donate (give) H^+ ions |

What citrus fruits does this photograph show? What acid is present in citrus fruit?

Name of acid	Formula	Ions produced from acid
Acetic acid	CH_3COOH	$H^+ + CH_3COO^-$
Carbonic acid	H_2CO_3	$2H^+ + CO_3^{2-}$
Hydrochloric acid	HCl	$H^+ + Cl^-$
Nitric acid	HNO_3	$H^+ + NO_3^-$
Sulphuric acid	H_2SO_4	$2H^+ + SO_4^{2-}$
Sulphurous acid	H_2SO_3	$2H^+ + SO_3^{2-}$

Table 1: Some acids and the ions which they form in solution

Concentrated and dilute acids

Concentrated acids contain a lot of acid in a small amount of water. (Concentrated sulphuric acid has 98% sulphuric acid and only 2% water.) Dilute acids have a small amount of acid in a lot of water. (Dilute sulphuric acid has about 10% sulphuric acid and 90% water.) Dilute acids usually have concentrations of 2.0 moles per dm^3 or less.

Strong and weak acids

When different acids with the same concentration in moles per dm^3 are tested with the same indicator, they give different pHs (figure 1).

The different pH values mean that some acids produce H$^+$ ions more readily than others. The results in figure 1 show that hydrochloric acid, nitric acid and sulphuric acid produce more H$^+$ ions than the other acids. We say that they **dissociate** (split up) into ions more easily. We call them **strong acids** and the others **weak acids**. Strong acids and weak acids also show a difference in their reactions with metals. Strong acids react much, much faster than weak acids.

We can show the difference between strong and weak acids in the way we write equations for their dissociation.

$$HCl(aq) \rightarrow H^+(aq) + Cl^-(aq)$$

$$CH_3COOH(aq) \rightleftharpoons H^+(aq) + CH_3COO^-(aq)$$

A single arrow, in the hydrochloric acid equation, shows that *all* the HCl has formed H$^+$ and Cl$^-$ ions in aqueous solution. The arrows in opposite directions for acetic acid show that some of the CH$_3$COOH molecules have formed H$^+$ and CH$_3$COO$^-$ ions. Most of the acetic acid remains as *undissociated* CH$_3$COOH molecules.

In aqueous solution:
- *strong acids are completely dissociated into ions,*
- *weak acids are only partly dissociated into ions.*

Notice the difference between the terms 'concentration' and 'strength'. **Concentration** tells us how much solute is dissolved in the solution, and we use the words 'concentrated' and 'dilute'. **Strength** tells us how much of the acid is dissociated into ions, and we use the words 'strong' and 'weak'.

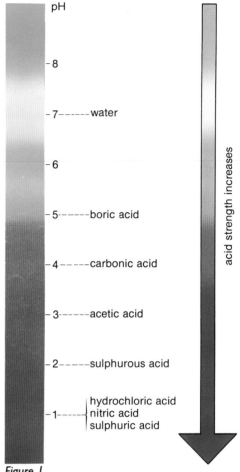

Figure 1
The pH of solutions of various acids (All solutions have a concentration of 0.1 mole per dm^3)

Questions

1 Make a table to summarize the properties and chemical reactions of acids.

2 (a) Why can vinegar (which contains acetic acid) be used to descale kettles?
(b) Why is sulphuric acid *not* used to descale kettles?
(c) Classify vinegar and battery acid (4 moles per dm^3 sulphuric acid as (i) concentrated or dilute, (ii) strong or weak.

3 Alkalis, like acids, can be strong or weak.
(a) sodium hydroxide is a strong alkali. Write an equation for its dissociation.
(b) How would you test to decide whether calcium hydroxide is a strong or weak alkali?

5 Sulphuric Acid

The manufacture of sulphuric acid was described in unit 1 of this section. About $2\frac{1}{2}$ million tonnes of sulphuric acid are manufactured each year in the UK. Figure 1 shows its main uses.

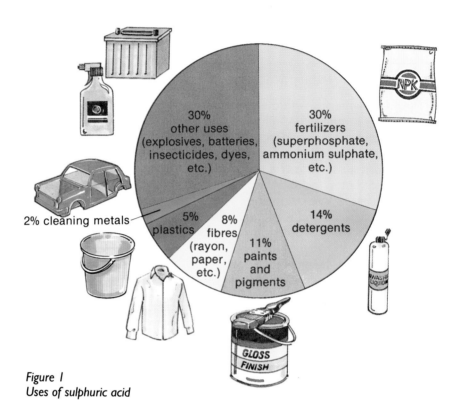

Figure 1
Uses of sulphuric acid

Concentrated sulphuric acid attacks skin, clothing, metals and other materials. Because of these hazards, containers for concentrated sulphuric acid must carry the 'corrosive' warning sign

Reaction with water

Concentrated sulphuric acid contains about 98% H_2SO_4 and only 2% water. **It must be handled carefully wearing eye protection.** It is an oily liquid which reacts with water producing a lot of heat. Because of this, **always add concentrated H_2SO_4 to a large volume of water when mixing the two. Never add water to acid.** If water is added to concentrated H_2SO_4, the heat produced can boil the water and spit out drops of acid.

Pure H_2SO_4 and pure water are both poor conductors of electricity. They contain simple molecules *not* ions. But, a solution of sulphuric acid in water is a good conductor of electricity, so it *must* contain ions. As we would expect, the solution of sulphuric acid is a strong acid fully dissociated into ions. The ions are formed when the H_2SO_4 reacts with water.

$$H_2SO_4(l) \xrightarrow{water} 2H^+(aq) + SO_4^{2-}(aq)$$

Notice that each molecule of H_2SO_4 can produce two H^+ ions.

Reactions as an acid

Always **wear eye protection** if you are handling sulphuric acid. Dilute sulphuric acid contains H^+ ions. So, it reacts like a typical acid (section G, unit 4) with

1 indicators

2 metals above copper in the reactivity series, forming a metal sulphate and hydrogen:

$$Mg + H_2SO_4 \rightarrow MgSO_4 + H_2$$

3 bases (metal oxides and hydroxides), forming a metal sulphate and water:

$$MgO + H_2SO_4 \rightarrow MgSO_4 + H_2O$$

4 metal carbonates, forming a metal sulphate, carbon dioxide and water:

$$MgCO_3 + H_2SO_4 \rightarrow MgSO_4 + CO_2 + H_2O$$

10 mins

Warm the sugar and concentrated sulphuric acid

Concentrated sulphuric acid removes water from sugar and leaves a black mass of carbon

Reactions as a dehydrating agent

Concentrated sulphuric acid reacts violently with water. It absorbs water very rapidly and can be used to dry gases. It is a **dehydrating agent**. The concentrated acid also removes water from hydrated salts such as blue copper(II) sulphate ($CuSO_4.5H_2O$), from carbohydrates such as sugar ($C_{12}H_{22}O_{11}$) and from compounds containing hydrogen and oxygen in clothes and skin. This is why it burns and chars clothing and skin. Dilute sulphuric acid does *not* react as a dehydrating agent in this way. When concentrated H_2SO_4 is added to sugar and warmed gently, the reaction gets very hot. The mixture froths up into a steaming black mass of carbon.

$$C_{12}H_{22}O_{11} \xrightarrow[\text{H}_2\text{SO}_4]{\text{concentrated}} \underset{\text{carbon}}{12C} + \underset{\substack{\text{water removed} \\ \text{by concentrated } H_2SO_4}}{11H_2O}$$

Sugar, like other carbohydrates, contains carbon plus hydrogen and oxygen atoms in the ratio 2:1 as in water. Hence the name *carbohydrate*. Our flesh also contains carbohydrates. Conc. H_2SO_4 removes the water from these carbohydrates.

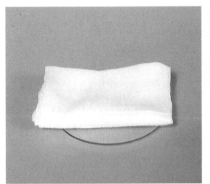 

Concentrated sulphuric acid will attack cotton cloth by removing water molecules from carbohydrates in the cotton fibres

Questions

1 (a) What is a dehydrating agent?
  (b) What uses do dehydrating agents have?
  (c) Accurate clocks sometimes have silica crystals (a dehydrating agent) placed near their working parts. Why is this?
2 (a) Write the formula for blue copper(II) sulphate crystals.
  (b) What colour will these crystals become when concentrated sulphuric acid is added?
  (c) Write an equation for the reaction.
3 Why does dilute sulphuric acid *not* react as a dehydrating agent?
4 (a) Why is concentrated sulphuric acid a poorer conductor of electricity than dilute sulphuric acid?
  (b) Why does concentrated sulphuric acid react with magnesium less vigorously than dilute sulphuric acid?
  (c) Why does concentrated sulphuric acid burn the skin but dilute sulphuric acid does not?
5 Which of the following are carbohydrates?
Ethanol, C_2H_6O; Glucose, $C_6H_{12}O_6$; Ethene, C_2H_4; Glycerine, $C_3H_8O_3$.

6 Bases and Alkalis

Bases are substances which neutralize acids. They are the chemical opposites to acids. The largest group of bases are metal oxides and metal hydroxides such as sodium oxide, copper oxide, sodium hydroxide, and copper hydroxide.

> *A special class of bases are called* **alkalis**. *Alkalis are bases which are soluble in water. Their solutions have a pH above 7. They turn litmus blue and give a green, blue or purple colour with Universal Indicator.*

The most common alkalis are sodium hydroxide (NaOH), calcium hydroxide ($Ca(OH)_2$) and ammonia (NH_3). Calcium hydroxide is much less soluble than sodium hydroxide. A solution of calcium hydroxide in water is often called 'limewater'. Sodium oxide (Na_2O), potassium oxide (K_2O) and calcium oxide (CaO) react with water to form their hydroxides. So the reaction of these three metal oxides with water produces alkalis. For example

$$Na_2O(s) + H_2O(l) \rightarrow 2NaOH(aq)$$

$$CaO(s) + H_2O(l) \rightarrow Ca(OH)_2(aq)$$

Most other metal oxides and hydroxides are insoluble in water. These insoluble metal oxides and hydroxides are bases but *not* alkalis. The relationship between bases and alkalis is shown in a Venn diagram in figure 1. All alkalis except ammonia solution are the hydroxides of reactive metals. They have similar properties because they all contain hydroxide ions (OH^-).

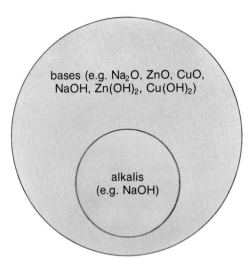

Figure 1
This Venn diagram shows the relationship between bases and alkalis. All alkalis are bases, but not all bases are alkalis

bases (e.g. Na_2O, ZnO, CuO, NaOH, $Zn(OH)_2$, $Cu(OH)_2$)

alkalis
(e.g. NaOH)

Explaining neutralization in terms of ions

Bases react with acids to form a salt and water. For example

copper oxide	+	sulphuric acid	→	copper sulphate	+	water
CuO(s)	+	$H_2SO_4(aq)$	→	$CuSO_4(aq)$	+	$H_2O(l)$

sodium hydroxide	+	hydrochloric acid	→	sodium chloride	+	water
NaOH(aq)	+	HCl(aq)	→	NaCl(aq)	+	$H_2O(l)$

These bases contain either oxide ions (O^{2-}) like copper oxide, or hydroxide ions (OH^-) like sodium hydroxide. During neutralization these ions react with H^+ ions in the acids to form water.

$$2H^+ + O^{2-} \rightarrow H_2O$$

$$H^+ + OH^- \rightarrow H_2O$$

These equations will help you to understand why bases are the chemical opposites to acids. Bases *take* H^+ ions, whereas acids *donate* them.

Alkalis in industry

The most important industrial alkalis are sodium hydroxide (caustic soda) and calcium hydroxide (slaked lime). Calcium hydroxide is made by adding water to quicklime (calcium oxide). It is used to treat acid soils, in cement and in the manufacture of sodium hydroxide and

> Always wear eye protection if you are using alkalis. Generally alkalis are more dangerous to the eyes than acids of the same concentration.

A 'noodling' machine used in part of the soap making process

bleaching powder. Large amounts of sodium hydroxide are used to make soap, paper, rayon and other cellulose fibres.

Paper, rayon and cellulose fibres are all made from wood. The wood is made into pulp and soaked in sodium hydroxide solution. This removes gums and resins and leaves the natural fibres of cellulose. The cellulose fibres are then squashed into thin white sheets which look like blotting paper. This purified pulp can now be used to make paper, rayon or cellulose acetate (Tricel).

Soaps and soap powders are made by boiling fats and oils with sodium hydroxide in large vats (figure 2).

In the manufacture of paper, wood pulp (cellulose) is purified by soaking in concentrated sodium hydroxide solution

oil or fat molecule   soap   glycerine (glycerol)

(/\/\/\/\— ≡ long chain of carbon, hydrogen and oxygen atoms)

Figure 2

Rayon is made by forcing a solution containing cellulose (wood pulp) through tiny holes into dilute acid to form filaments of yarn

Questions

1 Suppose your best friend has missed the last few lessons, and asks you for help. How would you explain to him or her what the difference is between (i) acids and bases, (ii) bases and alkalis, (iii) bases and salts?

2 Explain the following:
(a) All alkalis are bases, but all bases are *not* alkalis.
(b) Sodium hydroxide is used in oven cleaners to remove fat and grease.
(c) Limestone, slaked lime and lime are all bases, but only slaked lime is an alkali.

3 Write word equations and then balanced equations for the following reactions:
(a) the action of heat on limestone (calcium carbonate),
(b) the reaction of lime (calcium oxide) with water,
(c) the reaction of zinc oxide with hydrochloric acid.

4 (a) Why is it important to recycle paper?
(b) What are the main stages in recycling paper?
(c) What are the problems in recycling paper?
(d) Paper can be made from certain rags. Why is this?

7 Salts

Large cubic crystals of galena (lead sulphide)

Salts are formed when acids react with metals, bases or carbonates. Most salts contain a positive metal ion and a negative non-metal or radical ion. Salts are ionic compounds:

1 they have high melting points and boiling points,
2 they are electrolytes,
3 they are often soluble in water.

The best known salt is sodium chloride, NaCl, which is often called common salt. Many ores and minerals are composed of salts. These include chalk and limestone (calcium carbonate), gypsum (calcium sulphate) and iron pyrites (a mixture of copper sulphide and iron sulphide).

Salt crystals, like those of sodium chloride, are often formed by crystallization from aqueous solution. When this happens, water molecules sometimes form part of the crystal structure. This occurs in Epsom salts ($MgSO_4.7H_2O$), gypsum ($CaSO_4.2H_2O$) and washing soda ($Na_2CO_3.10H_2O$). The water which forms part of the crystal structure is called **water of crystallization**. Salts containing water of crystallization are called **hydrates** or hydrated salts.

Soluble and insoluble salts

If you are using a salt or making a salt, it is important to know whether it is soluble or insoluble.

Table 1 shows the solubilities of various salts in water at 20° C. Notice the wide range in solubilities from potassium nitrite (300 g per 100 g water) to silver chloride (0.000 000 1 g per 100 g water). It is useful to divide salts into two categories—soluble and insoluble.

Purple cubic crystals of fluorite (calcium fluoride)

Salt	Formula	Solubility /g per 100 g water at 20°C
Barium chloride	$BaCl_2$	36.0
Barium sulphate	$BaSO_4$	0.000 24
Calcium chloride	$CaCl_2$	74.0
Calcium sulphate	$CaSO_4$	0.21
Copper(II) sulphate	$CuSO_4$	20.5
Copper(II) sulphide	CuS	0.000 03
Lead(II) sulphate	$PbSO_4$	0.004
Potassium chlorate	$KClO_3$	7.3
Potassium nitrite	KNO_2	300.0
Silver chloride	$AgCl$	0.000 000 1
Silver nitrate	$AgNO_3$	217.0
Sodium chloride	$NaCl$	36.0
Sodium nitrate	$NaNO_3$	87.0

Table 1: The solubilities of various salts

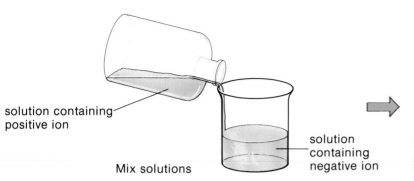

solution containing
positive ion

Mix solutions

solution
containing
negative ion

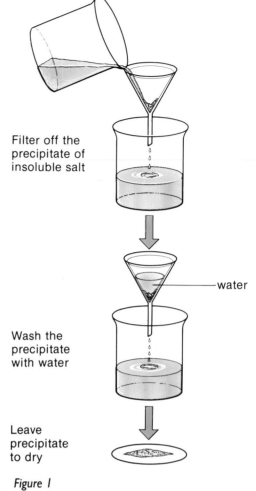

Filter off the
precipitate of
insoluble salt

water

Wash the
precipitate
with water

Leave
precipitate
to dry

Figure 1
Preparing an insoluble salt

Salts with a solubility greater than 1 g per 100 g water are classed as soluble; salts with a solubility less than 1 g per 100 g water are classed as insoluble.

Table 2 summarizes the general rules for the solubilities of common salts.

All	sodium potassium ammonium	salts are soluble	
All	nitrates	are soluble	
All	sulphates	are soluble except	Ag$_2$SO$_4$ BaSO$_4$ CaSO$_4$ PbSO$_4$
All	chlorides	are soluble except	AgCl PbCl$_2$
All	carbonates sulphides sulphites	are **insoluble** except	those of Na$^+$, K$^+$ and NH$_4$$^+$

Table 2: Solubilities of common salts

Preparing insoluble salts

The method used to prepare a salt depends on whether the salt is soluble or insoluble. Methods for soluble salts are described in the next unit. Insoluble salts, like lead chloride, silver chloride, calcium carbonate and barium sulphate, are prepared by making the salt as a precipitate.

Suppose you are making insoluble silver chloride, AgCl. You will need to mix a soluble Ag$^+$ salt and a soluble chloride.

1 Which Ag$^+$ salt is certain to be soluble? Look at table 2.
2 Which chloride is certain to be soluble? Look at table 2.

These questions show that you can precipitate any insoluble salt (say XY) by mixing solutions of NaY and XNO$_3$. Both NaY and XNO$_3$ are soluble, since all sodium salts and all nitrates are soluble. Figure 1 shows how an insoluble salt is precipitated and then purified. Make one solution containing the positive ion in the insoluble salt and another solution containing the negative ion. Mix the two solutions, filter off the insoluble salt, wash it with water and then leave it to dry at room temperature.

Questions

1 Explain the following:
hydrated; *water of crystallization*; *precipitation*; *insoluble*.
2 What units are used for (i) concentration; (ii) solubility?
3 (a) Summarize the stages in preparing an insoluble salt.
  (b) Describe how you would prepare a pure sample of insoluble barium sulphate.
  (c) Write an equation for the reaction which occurs.
4 Make a table to show whether the following salts are soluble or insoluble:
Pb(NO$_3$)$_2$; Ag$_2$S; CuCO$_3$; K$_2$SO$_3$; NH$_4$Cl; FeSO$_4$.
5 Epsom Salts have the formula MgSO$_4$.7H$_2$O. What does this tell you about Epsom Salts?

8 Preparing Salts

When you are making a salt, the first question to ask is 'Is the salt soluble or insoluble?'. If the salt is insoluble, it is usually prepared by precipitation (see unit 7 of this section).

Soluble salts are usually prepared by reacting an acid with a metal, a base or a carbonate (see unit 4 of this section).

$$\text{metal} + \text{acid} \rightarrow \text{salt} + H_2$$
$$\text{base} + \text{acid} \rightarrow \text{salt} + H_2O$$
$$\text{carbonate} + \text{acid} \rightarrow \text{salt} + CO_2 + H_2O$$

● **Method 1: Using metals, insoluble bases and insoluble carbonates.** Figure 1 shows the main stages in this method.

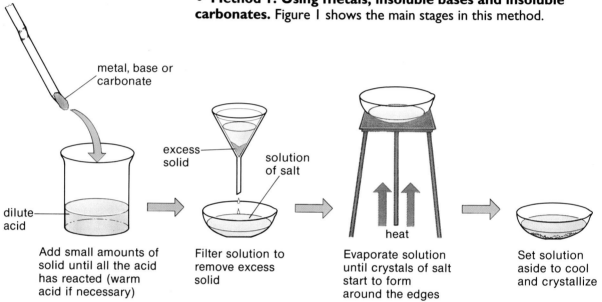

metal, base or carbonate

excess solid

solution of salt

dilute acid

heat

Add small amounts of solid until all the acid has reacted (warm acid if necessary)

Filter solution to remove excess solid

Evaporate solution until crystals of salt start to form around the edges

Set solution aside to cool and crystallize

Figure 1

Zinc sulphate can be made by this method using sulphuric acid with either zinc, zinc oxide or zinc carbonate.

$$Zn + H_2SO_4 \rightarrow ZnSO_4 + H_2$$
$$ZnO + H_2SO_4 \rightarrow ZnSO_4 + H_2O$$
$$ZnCO_3 + H_2SO_4 \rightarrow ZnSO_4 + CO_2 + H_2O$$

● **Method 2: Using soluble bases and carbonates.** In method 1, we can tell when the acid has been used up because unreacted metal, base or carbonate remains in the liquid as undissolved solid. But, if the base or carbonate is soluble (like sodium hydroxide or sodium carbonate), we cannot tell when the acid has been used up, because excess solid will dissolve even after the acid has been neutralized. To get round this, we must use an indicator to tell us when we have added just enough base or carbonate to neutralize the acid. Figure 2 shows the main stages involved.

Potassium chloride can be made by this method using hydrochloric acid with either potassium hydroxide or potassium carbonate.

$$KOH + HCl \rightarrow KCl + H_2O$$
$$K_2CO_3 + 2HCl \rightarrow 2KCl + CO_2 + H_2O$$

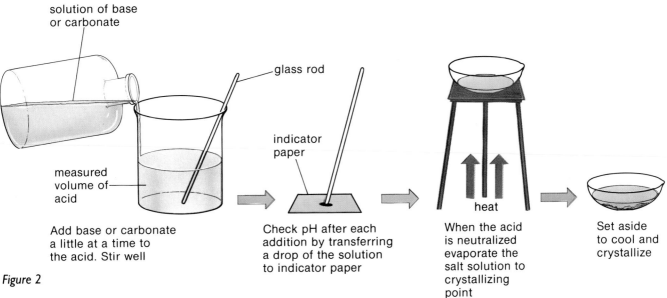

solution of base
or carbonate

glass rod

indicator
paper

heat

measured
volume of
acid

Add base or carbonate
a little at a time to
the acid. Stir well

Check pH after each
addition by transferring
a drop of the solution
to indicator paper

When the acid
is neutralized
evaporate the
salt solution to
crystallizing
point

Set aside
to cool and
crystallize

Figure 2

Method 2 is used to make the salts of sodium, potassium and ammonium, because the bases and carbonates containing sodium, potassium and ammonium are all soluble. Other soluble salts are usually made by method 1. Figure 3 shows a flowchart which can be used to decide how to prepare a particular salt.

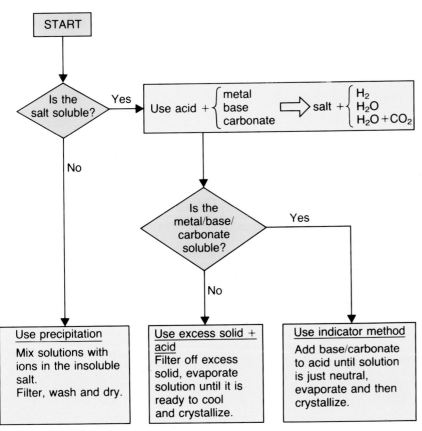

START

Is the
salt soluble?

Yes

Use acid + { metal, base, carbonate } $\Rightarrow$ salt + { H_2, H_2O, $H_2O + CO_2$ }

No

Is the
metal/base/
carbonate
soluble?

Yes

No

Use precipitation

Mix solutions with
ions in the insoluble
salt.
Filter, wash and dry.

**Use excess solid +
acid**

Filter off excess
solid, evaporate
solution until it is
ready to cool
and crystallize.

Use indicator method

Add base/carbonate
to acid until solution
is just neutral,
evaporate and then
crystallize.

Figure 3
How to prepare a salt

Questions

1 Look at method 1 for preparing soluble salts.

(a) Explain why this method will not work for metals below hydrogen in the activity series.

(b) Why is this method not used with sodium?

(c) How can you tell when all the acid is used up if the solid used is (i) zinc; (ii) copper oxide; (iii) copper carbonate?

(d) Why is the salt produced not contaminated with (i) the acid used; (ii) the solid added?

(e) Why is method 1 no good for insoluble salts?

(f) Why is method 1 no good if the solid added dissolves in water?

2 Look at method 2 for preparing salts.

(a) Why is the pH of the solution tested using indicator paper rather than putting indicator solution into the acid?

(b) Describe how you would make sodium nitrate by this method.

(c) Write a word equation and a balanced symbolic equation for the reaction in (b).

9 Lime and Limestone

A limestone quarry. The mining of limestone is big business because it has so many commercially valuable products

Limestone for industry

Limestone is one of the most important raw materials for the chemical and building industries. The main substance in limestone rock is calcium carbonate ($CaCO_3$). Quicklime (calcium oxide, CaO) and slaked lime (calcium hydroxide, $Ca(OH)_2$) are easily manufactured from limestone, which makes it even more important. The uses of limestone and its products are shown in figure 1. Notice:

1 the main uses of limestone itself are in making iron and steel (section F, unit 6), in neutralizing acid soil and in fertilizers such as nitro-chalk.

2 the main substances made from limestone are calcium oxide (quicklime), calcium hydroxide (slaked lime), cement and glass.

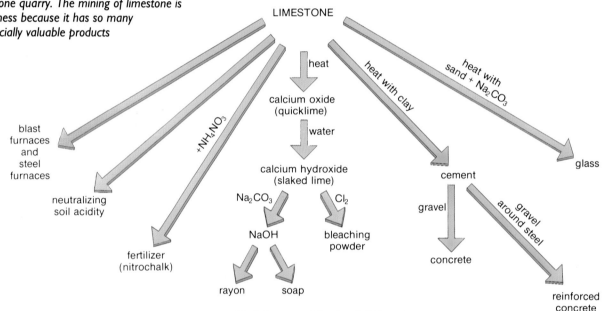

Figure 1
Important uses and products of limestone

• **Calcium oxide (quicklime)** is made by heating limestone in lime kilns (figure 2). Calcium carbonate in the limestone is decomposed to calcium oxide (quicklime).

$$CaCO_3(s) \rightarrow CaO(s) + CO_2(g)$$

The quicklime is then reacted with water to make calcium hydroxide (slaked lime).

$$CaO(s) + H_2O(l) \rightarrow Ca(OH)_2(s)$$

• **Calcium hydroxide (slaked lime)** is the cheapest industrial alkali. It is used to make sodium hydroxide and bleaching powder. Powdered calcium hydroxide is also used to control acidity in the soil. Calcium hydroxide is only slightly soluble in water. The dilute alkaline solution which it forms is called **limewater**.

• **Cement** is made by heating limestone with clay. It contains a mixture of calcium silicate and aluminium silicate. This mixture reacts with water to form hard, interlocking crystals as the cement sets. When cement is used, it is normally mixed with two or three times as much sand as well as water.

- **Concrete** is a mixture of cement, sand and water with gravel (aggregate), broken stones or bricks.

Reinforced concrete is made by allowing concrete to set around a steel framework. It is used in building large structures like office blocks and bridges. Reinforced concrete is an example of a **composite material**. Composite materials contain two different materials, but the particles in the materials do not mix. Reinforced concrete contains concrete and steel. Because of this, it has the hardness of concrete and the flexibility and tensile strength of steel.

Chalk, limestone and marble

Calcium carbonate occurs naturally as chalk and marble, as well as limestone. In fact, calcium carbonate is the second most abundant material in the Earth's crust after silicates like clay, sand and sandstone.

Chalk is the softest form of calcium carbonate. Deposits of chalk have formed from the shells of dead sea creatures that lived millions of years ago. In some places, the chalk was covered with other rocks and put under great pressure. This changed the soft chalk into a harder rock—limestone. In other places the chalk was under pressure *and heat*. This changed the soft chalk into marble, the hardest form of calcium carbonate.

Reinforced concrete being used in building a bridge

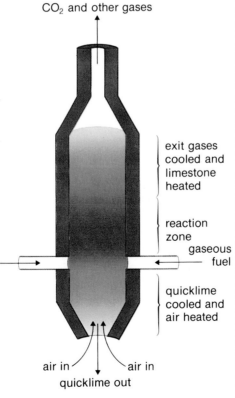

CO$_2$ and other gases

exit gases cooled and limestone heated

reaction zone

gaseous fuel

quicklime cooled and air heated

air in   air in
quicklime out

Figure 2
A gas-fuelled lime kiln

Questions

1 Chalk, marble and limestone are different natural forms of calcium carbonate.
   (a) Why are there three different forms?
   (b) Find out about the classification of rocks as 'sedimentary', 'metamorphic' and 'igneous'. Which classes of rocks do chalk, marble and limestone belong to?

2 Various mixtures of sand and cement (by volume) are used in building. Plan an experiment to find out which mixture (ratio of sand to cement 2:1, 2.5:1, 3:1 or 3.5:1) gives the *hardest* product.

3 Try to explain the following. Write balanced equations where appropriate.
   (a) Finely ground limestone is used to neutralize acids in the soil.
   (b) Lime water is used to test for carbon dioxide.

4 'Wattle and daub' might be described as the medieval equivalent of reinforced concrete as a composite material. Find out about 'wattle and daub'. Why is it a composite material? What was the advantage of 'wattle and daub' over just 'wattle' or just 'daub'?

5 (a) What problems are associated with the large-scale quarrying of chalk and limestone?
   (b) What steps can be taken to overcome or reduce these problems?

10 Hard Water

This photo is of Malham Cove in the Yorkshire Dales. Water once poured from the top of the rocks. Over many years carbonic acid in the water has reacted with the limestone and formed caves. The water now flows through these caves and finally emerges at the bottom of the cove

In chalk and limestone areas, soap does not lather easily with the water. The water also forms a **scum** when it is mixed with soap. This kind of water is known as **hard water**.

Which ions cause hard water?

The table shows what happens when 10 cm³ of each of the solutions listed are shaken with 1.0 cm³ of soap solution.

Solution used	Ions present	Reaction with hard water
Sodium chloride	Na^+, Cl^-	No scum, lots of lather
Calcium chloride	Ca^{2+}, Cl^-	Lots of scum, little lather
Potassium nitrate	K^+, NO_3^-	No scum, lots of lather
Magnesium nitrate	Mg^{2+}, NO_3^-	Lots of scum, little lather
Sodium sulphate	Na^+, SO_4^{2-}	No scum, lots of lather
Iron(II) sulphate	Fe^{2+}, SO_4^{2-}	Lots of scum, little lather

The reactions of some solutions with soap

Just look at the scum which has formed with this hard water. Notice also that there is no lather

Look closely at the table.
1 Which solutions cause hard water?
2 Which of the following ions cause hard water: Na^+; Cl^-; Ca^{2+}; K^+; NO_3^-; Mg^{2+}; SO_4^{2-}; Fe^{2+}?
3 Which ions are most likely to cause hard water in the UK?

Why does scum form?

The main cause of hard water in the UK are calcium ions, Ca^{2+}. Soaps contain salts such as sodium palmitate and sodium stearate. When hard water is mixed with soap, Ca^{2+} ions in the hard water react with palmitate and stearate ions in the soap forming an insoluble precipitate of calcium stearate and calcium palmitate. This precipitate is scum.

$$Ca^{2+}(aq) \quad + \quad 2X^-(aq) \quad \rightarrow CaX_2(s)$$

calcium ions   stearate/palmitate   scum
in hard water   ions in soap

Soaps and detergents are both used for cleaning, but they differ in *one* important way. Detergents, like washing-up liquid, do *not* give a scum with hard water. Unlike soaps they do not contain ions which react with Ca^{2+} ions in hard water to form a precipitate.

How does hard water form?

When rain falls, it reacts with carbon dioxide in the air to form carbonic acid.

$$H_2O(l) + CO_2(g) \rightarrow H_2CO_3(aq)$$

When this dilute solution of carbonic acid flows over limestone or chalk it reacts with calcium carbonate in the rocks to form calcium hydrogencarbonate.

$$CaCO_3(s) \quad + \quad H_2CO_3(aq) \quad \rightarrow \quad Ca(HCO_3)_2(aq)$$

in limestone   in rain water   in hard water

Unlike calcium carbonate, calcium hydrogencarbonate is soluble in water and the calcium ions make the water hard.

Calcium carbonate in chalk and limestone is the main cause of hard water. In some areas, calcium sulphate which occurs as gypsum ($CaSO_4.2H_2O$) and anhydrite ($CaSO_4$) also causes hardness. Calcium sulphate is only slightly soluble in water but enough will dissolve to make the water hard.

In limestone areas, caves have formed as carbonic acid in rain water has reacted with the limestone rock. These caves are at Castleton in Derbyshire

Carbonic acid in rain water has helped to dissolve the limestone in the cracks of this limestone pavement

Questions

1 Explain the following:
hard water; scum; soap; detergent.
2 (a) Why does scum form?
   (b) Write an equation for the reaction involved.
   (c) Why is scum a nuisance?
3 What is the main advantage of detergents over soaps?
4 Plan an experiment to compare the hardness of two different samples of tap water.
5 Design and make a poster to show some of the geological features in limestone areas that have resulted from the effects of dissolved carbon dioxide on limestone.

11 Softening Hard Water

'Fur' inside a kettle is a deposit of calcium carbonate from hard water

Hard water usually tastes better than soft water. The dissolved substances in hard water also help to produce strong teeth and bones which contain calcium carbonate and calcium phosphate. But hard water has several disadvantages compared to soft water.

1 It uses more soap than soft water.

2 It produces scum, which looks unsightly.

3 It results in the formation of 'scale' in water pipes and 'fur' in kettles. The 'scale' may block pipes and 'fur' will reduce the efficiency of a kettle.

4 It is necessary to remove the hardness from the water in some areas and this causes extra expense.

When hard water is warmed up or boiled, the calcium hydrogencarbonate in it is decomposed to calcium carbonate, water and carbon dioxide.

$$Ca(HCO_3)_2(aq) \rightarrow CaCO_3(s) + H_2O(l) + CO_2(g)$$

The calcium carbonate is insoluble and forms a deposit inside the pipe or kettle. This reaction is the reverse of that which forms hard water. The reaction also explains the formation of stalagmites and stalactites in limestone caves. The temperature inside the cave is just warm enough for some of the hard water to decompose and leave a tiny deposit of calcium carbonate. More water drips down and the deposit gets larger. A deposit also forms where the drops hit the floor. After hundreds of years, the deposits will grow into large stalagmites and stalactites.

Stalagmites and stalactites in Cox's Cave, Cheddar, Somerset

How is hard water softened?

In some areas, substances that cause hardness in the water must be removed. This is called **water softening**. In order to soften hard water we must take out the Ca^{2+} ions. We can do this in various ways.

- **By boiling.** Boiling decomposes calcium hydrogencarbonate forming insoluble calcium carbonate.

$$Ca(HCO_3)_2(aq) \rightarrow CaCO_3(s) + H_2O(l) + CO_2(g)$$

This removes the hardness caused by calcium hydrogencarbonate, but boiling does not remove the hardness caused by calcium sulphate. Because of this, the hardness from calcium sulphate is called **permanent hardness**. The hardness caused by calcium hydrogencarbonate (which is removed by boiling) is called **temporary hardness**.

- **By distillation.** Distillation produces pure water, removing both permanent and temporary hardness.

- **By adding washing soda.** Washing soda and bath salts contain sodium carbonate (Na_2CO_3). Adding these to hard water removes all the Ca^{2+} ions as a precipitate of calcium carbonate.

$$\underset{\text{in hard water}}{Ca^{2+}(aq)} + \underset{\text{in washing soda}}{CO_3^{2-}(aq)} \rightarrow CaCO_3(s)$$

- **By ion-exchange.** The most convenient way of softening water is to use an ion-exchange column. The water passes through a column containing a special substance called a **resin** (figure 1). The resin contains sodium ions which are displaced by calcium ions as hard water passes through the column. Na^+ ions do not cause hardness, so the water is now 'soft'.

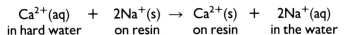

$$\underset{\text{in hard water}}{Ca^{2+}(aq)} + \underset{\text{on resin}}{2Na^+(s)} \rightarrow \underset{\text{on resin}}{Ca^{2+}(s)} + \underset{\text{in the water}}{2Na^+(aq)}$$

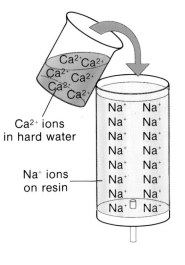

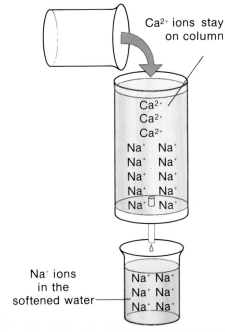

Figure 1
Using an ion-exchange column

Questions

1 Explain the following:
water softening; permanent hardness; temporary hardness; ion-exchange resin.

2 (a) What are the advantages of hard water?
(b) What are the disadvantages of hard water?

3 (a) How do stalagmites and stalactites form?
(b) Write an equation for the reaction involved in (a).

4 (a) List the main methods of softening water.
(b) For each method write an equation to summarize the reaction involved.

5 Water in London is much harder than water in Manchester. Why is this?

6 Limescale often forms on taps in hard water areas. Usually, there is more scale on the hot tap than on the cold tap.
(a) How does limescale form?
(b) Why is there usually more on the hot tap?

7 During the 1950s, tooth decay was worse in the North West of England than in the South East. Why was this?

Section G: Activities

1 | Making fertilizers

The flow diagram below shows the raw materials which are used to make the two most important fertilizers:

- N fertilizer ('straight' fertilizer) containing ammonium nitrate,
- NPK fertilizer ('compound' fertilizer) containing ammonium nitrate, ammonium phosphate and potassium chloride.

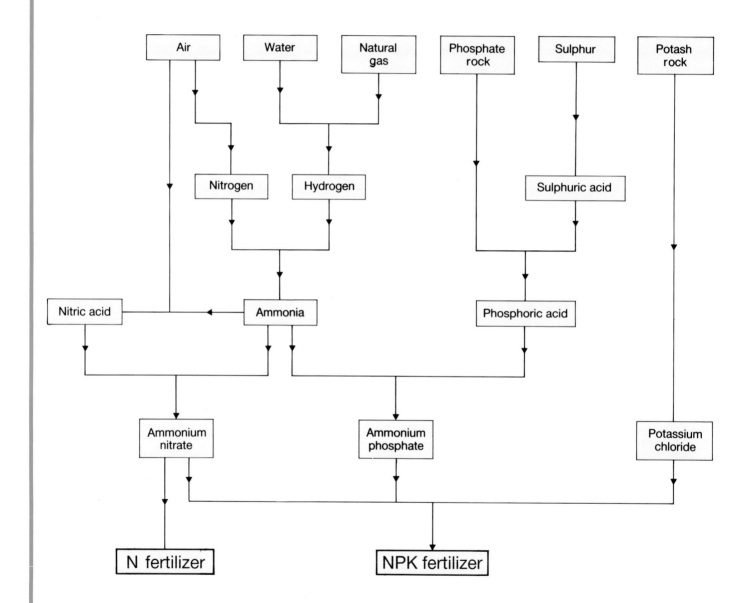

1 What raw materials besides air and water are used to make these fertilizers?

2 Where do the raw materials in question 1 come from to manufacture these fertilizers in the UK?

3 The reaction of ammonia (NH_3) with phosphoric acid (H_3PO_4) to form ammonium phosphate involves neutralization.
(a) What is neutralization?
(b) Write a word equation for this reaction.
(c) Use chemical symbols to write a balanced equation for the reaction.
4 One other reaction in the flow diagram also involves neutralization. Write a word equation and a balanced symbolic equation for this reaction too.
5 Phosphoric acid is manufactured from phosphate rock which contains calcium phosphate ($Ca_3(PO_4)_2$).
The equation for the reaction is

$$Ca_3(PO_4)_2 + 3H_2SO_4 \rightarrow 2H_3PO_4 + 3CaSO_4$$

$$(H = 1, O = 16, P = 31, Ca = 40)$$

(a) What is the relative formula mass of calcium phosphate?
(b) What is the relative formula mass of phosphoric acid?
(c) According to the equation above, how many grams of phosphoric acid are produced from 310 g of calcium phosphate?
(d) How much calcium phosphate is needed to produce 100 tonnes of phosphoric acid?

Most of the ammonia produced at this plant is used to make fertilizers

2 | pH changes after eating a sweet

Using a special instrument called a pH meter, Dr Razell measured the pH in Lee's mouth after he had eaten a sugary sweet.
The results are shown in table 1.
1 Open up a new spreadsheet on your computer.
2 Enter the times and pHs as in the table above.
3 From the spreadsheet, plot a line graph of pH (y-axis) against time (x-axis). (If you cannot plot graphs directly from the spreadsheet, draw the graph by hand.)
4 Print a copy of your line graph.
Sugary foods are converted to acids in our mouths. These acids attack the enamel on our teeth and cause tooth decay. Tooth decay occurs when the pH is lower than 5.5.
5 What does pH measure?
6 Describe how the pH changes in Lee's mouth after eating the sweet.
7 How many minutes after eating the sweet does tooth decay start?
8 How many minutes after eating the sweet does tooth decay stop?
9 How many minutes after eating the sweet is tooth decay at its worst? Explain your answer.
10 Use your graph to explain why dentists advise against eating sweets between meals.
11 What should the pH of toothpaste be? Explain your answer.
12 Dental decay can be prevented by strengthening the enamel of teeth. This is often done using a chemical in drinking water or in toothpaste. Find out more about this.

Time after eating sweet/minutes	pH
0	6.6
2	5.6
4	5.2
6	5.0
8	4.9
10	4.9
12	5.1
14	5.3
16	5.5
18	5.6
20	5.8

Table 1: How pH changed in Lee's mouth after eating a sweet

Section G: Study Questions

1 Question 1 concerns acids.
(a) Name *one* indicator that you could use to test for an acid.
(b) What would happen to the indicator that you named in part (a) when you placed it in an acidic solution?
(c) Name the acid that is used in a car battery.
(d) Name *one* substance, usually found around the house, that you could use to neutralize some spilt battery acid.
(e) Bath salts contain crystals of sodium carbonate. What gas would be given off if vinegar (ethanoic acid) were dropped onto some of these crystals?

2 The pH scale is used to indicate how acidic or alkaline a solution is. Here are some numbers from the pH scale:
      1   3   7   9   14
(a) What kind of a solution would have a pH of 1?
(b) Which pH number from the above list would lime water have?
(c) Which number represents the pH of pure water?

3 Complete the table below, which describes the preparation of some salts.

REACTANTS	PRODUCTS
magnesium oxide + ? →	magnesium sulphate + ?
? + ? →	zinc chloride + hydrogen

4 Magnesium sulphate crystals ($MgSO_4.7H_2O$) can be made by adding excess magnesium oxide (MgO), which is insoluble in water, to dilute sulphuric acid.
(a) Why is the magnesium oxide added in excess?
(b) The following apparatus could be used to separate the excess magnesium oxide from the solution. What words should be used for the labels A to D?

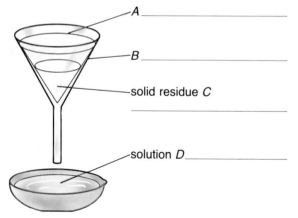

A _____

B _____

solid residue C

_____

solution D _____

(c) After the excess magnesium oxide has been removed, the solution is partly evaporated and set aside. Some days later the sample is filtered again and the product is washed and finally dried.
(i) Why is the solution partly evaporated?
(ii) Which substance is removed by the *second* filtering?
(iii) How could the product be dried?
(iv) What would happen to the product if it was then heated strongly?

5 (a) Three unlabelled bottles are known to contain soft water, permanently hard water and temporarily hard water. Describe how you would distinguish between these different samples of water using the following tests.
Test 1: a small amount of soap was added to each sample of water which was then shaken.
Test 2: each sample of water was boiled, allowed to cool, and a small amount of soap was added.
(b) The water supply to a house passes through a container labelled *'Ion Exchange Resin'*.
(i) What does this resin do to the water?
(ii) Name an ion removed from the water by this resin.
(iii) Give one disadvantage of treating domestic water supplies in this way.
(iv) How do phosphates get into the water system?
(v) What would be the effect of lowering the amount of dissolved oxygen in streams and lakes?

6 Study the following reaction scheme:

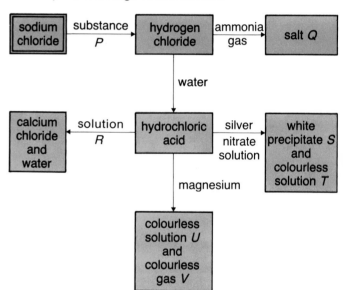

(a) Give the name of
substance P   solution T
salt Q   solution U
solution R   gas V
precipitate S
(b) (i) State the type of reaction taking place when hydrochloric acid reacts with solution R.
(ii) Name one other compound that will give a white precipitate with silver nitrate solution.
(iii) Describe a test by which you could identify gas V.

SECTION H

The Structure of Materials

Mica has a structure in which the particles are arranged in layers. These layers can be separated very easily

1 Studying Structures

Crystals of impure rock salt. Even in this impure sample you can see the cubic shape of sodium chloride crystals

White quartz crystals on impure dolomite. Quartz is silicon dioxide. Dolomite is a mixture of calcium carbonate and magnesium carbonate

Look at the crystals of rock salt and quartz in the photographs above. What do you notice about all the salt crystals? What do you notice about all the quartz crystals? All the salt crystals are roughly the same cubic shape. All the quartz crystals are roughly cylindrical with pointed tops. Further studies show that *all the crystals of one substance have similar shapes*. This suggests that the particles in the crystal are always packed in a regular fashion to give the same overall shape. Sometimes, crystals grow unevenly and their shapes become distorted. Even so, it is usually easy to see their general shape. Solid substances which have a regular packing of particles are described as **crystalline**. The particles may be atoms, ions or molecules. Figure 1 shows how cubic crystals and hexagonal crystals can form. If the particles are always placed in parallel lines or at 90° to each other, the crystal will be cubic. If the particles are placed at 120° in the shape of a hexagon, the final crystal will be hexagonal.

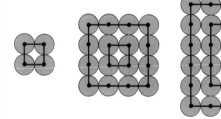

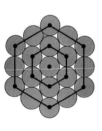

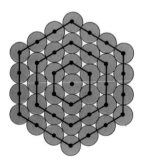

When the particles are arranged in a cubic fashion the final crystal will be cubic

When the particles are arranged in a hexagonal fashion the final crystal will be hexagonal

Figure 1

We can compare the way in which a crystal grows to the way in which a bricklayer lays bricks. If the bricklayer always places the bricks in parallel lines or at 90° to each other, then the final buildings will be like cubes or boxes. However, if some bricks are laid at 120° to make hexagons, then the final buildings will be hexagonal.

The overall shape of a crystal can only give a *clue* to the way in which the particles are arranged. X-rays give much *better evidence*.

Using X-rays to study crystals

Look through a piece of thin stretched cloth at a small bright light. The pattern you see is due to the deflection of the light as it passes through the regularly spaced threads of the fabric. This deflection of light is called **diffraction** and the patterns produced are **diffraction patterns**. If the cloth is stretched so that the threads in the fabric get closer, then the pattern spreads further out. From the diffraction pattern which we *can* see, we can work out the pattern of the threads in the fabric which we *cannot* see. The same idea is used to work out how the particles are arranged in a crystal.

A narrow beam of X-rays is directed at a well-formed crystal (figure 2). Some of the X-rays are diffracted by particles in the crystal onto X-ray film. When the film is developed, a regular pattern of spots appears. This is the diffraction pattern for the crystal. From the diffraction pattern which we *can* see, it is possible to work out the pattern of particles in the crystal which we *cannot* see. A regular arrangement of spots on the film indicates a regular arrangement of particles in the crystal. This regular arrangement of particles in the crystal is called a **lattice**.

X-rays have been used to study thousands of different crystals. An X-ray diffraction photograph is shown in figure 3. Nowadays, we can also use rays of electrons, like X-rays, to study the way in which particles are arranged in crystals.

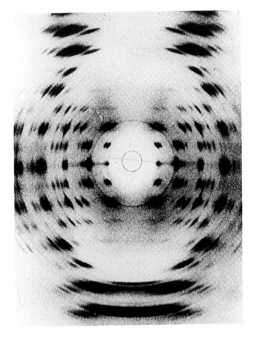

Figure 3
An X-ray diffraction photograph of DNA crystals. What general pattern do the dots form?

Snowflake crystals

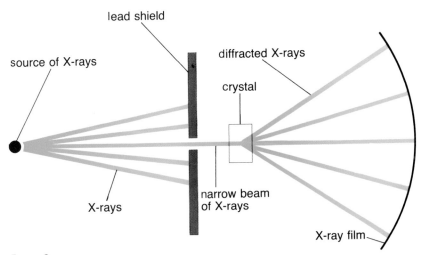

Figure 2
Diffraction of X-rays by a crystal

Labels in figure 2: lead shield; diffracted X-rays; crystal; source of X-rays; X-rays; narrow beam of X-rays; X-ray film

Questions

1 Explain the words:
crystal; lattice; diffraction.
2 Look at the snowflake crystals in the photograph above.
 (a) What substance makes up snowflakes?
 (b) What particles do snowflakes contain?
 (c) How do you think the particles are arranged in snowflakes?
3 How are X-rays used to give evidence for the arrangement of particles in a crystal?
4 Why do all crystals of one substance have roughly the same shape?

2 The Structure of Substances

What properties must the material in nappies have? What do you think the structure of nappy material is like?

Why was metal used to make suits of armour in the Middle Ages?

What properties of clay make it useful for making pots and crockery? What do you think the structure of clay is like?

The uses of materials depend on their properties. For example, copper is used for electrical wires and cables because it can be drawn into wires and it is a good conductor of electricity. All substances are made up of particles. If we know how these particles are arranged (the **structure**) and how the particles are held together (the **bonding**), then we can explain the **properties** of substances. For example, copper is a good conductor because its metallic bonding allows electrons to move through the structure when it is connected to a battery. It can be drawn into wires because copper atoms can slide over each other in the close-packed structure.

> *Notice that the structure and bonding of a substance determine its properties and, in turn, the properties determine its uses.*

The links from structure and bonding help us to explain why metals are used as conductors, why graphite is used in pencils and why clay is used to make bricks.

Using X-ray analysis, we can find out how the particles are arranged in a substance (its structure), but it is more difficult to study the forces between these particles (its bonding).

From previous sections, you will know that

> *all substances are made up of only three different types of particle—atoms, ions and molecules.*

These three particles give rise to four different solid structures.

● metallic structures

● simple molecular structures

● giant molecular structures

● ionic structures

Table 1 shows the particles in these structures, the types of substances formed and examples of these substances.

A large natural diamond embedded in an ore, side by side with a cut diamond. How do the structure and properties of diamond lead to its use in jewellery?

Type of structure	Particles in the structure	Types of substance	Examples
Metallic (see also section F)	atoms	metals	Na, Fe, Cu
Simple molecular	small molecules containing a few atoms	non-metals, or non-metal compounds	I_2 (iodine) O_2 (oxygen) H_2O (water) CO_2 (carbon dioxide)
Giant molecular	very large molecules containing thousands of atoms	non-metals, or non-metal compounds	diamond, graphite, polythene, sand
Ionic	ions	compounds of metals with non-metals	Na^+Cl^- (salt) $Ca^{2+}O^{2-}$ (lime)

Table 1: The four types of solid structure and the particles they contain

Lead sulphide (galena) crystals on a sample of iron(II) carbonate (siderite). What particles will lead sulphide contain? How do you think the particles are arranged?

Questions

1 Get into a small group with 2 or 3 others. Look at the photos in this unit and discuss the questions in the captions.

2 (a) What are the particles in metal structures?

(b) Why are most metal structures described as close-packed?

(c) How are the particles arranged in most metal structures?

3 What type of structure will the following substances have?
chlorine, limestone (calcium carbonate), silver, air, rubber, polyvinylchloride (PVC), brass, wood.

4 Conductivity tests can give evidence for the particles, bonding and structure of substances. Describe (i) the tests you would carry out, (ii) the results you would expect, (iii) the conclusions you would make from your results. (Hint: See section D, units 3 and 4.)

3 Carbon: Diamond

Diamond, graphite and charcoal are all made of pure carbon. But these three solids have very different properties and uses. Diamond is hard and clear, whereas graphite and charcoal are soft and black. Diamonds are used to cut and engrave glass, but graphite and charcoal are used by artists to get a soft, shaded effect.

An artist using a stick of charcoal

This glass engraving wheel has been toughened with diamond. Diamond is the best material to use to toughen the wheel because it is the hardest natural material

These different forms of solid carbon are called **allotropes**. A few other elements also have allotropes. Oxygen has two allotropes—oxygen (O_2) and ozone (O_3). Sulphur has three allotropes—rhombic sulphur, monoclinic sulphur and plastic sulphur. *Allotropes are different forms of the same element in the same state.*

Diamond, graphite and charcoal have different properties and different uses because *they have different structures.* Their atoms are not packed in the same way.

The arrangement of carbon atoms in diamond, graphite and charcoal has been studied by X-ray analysis.

Diamond

In diamond, each carbon atom is joined to four other atoms (figure 1). Each atom is at the centre of a tetrahedron surrounded by four others at the corners of the tetrahedron (figure 2). Every carbon atom shares electrons with each of its four neighbours forming strong covalent bonds. The covalent bonds extend through the whole diamond, forming a three-dimensional structure. Thus, a diamond is a single **giant molecule** or **macromolecule**.

Only a small number of atoms are shown in the model in figure 1. In a real diamond, this arrangement of carbon atoms is repeated millions and millions of times.

Figure 1
An 'open' model of the diamond structure

Properties and uses of diamond

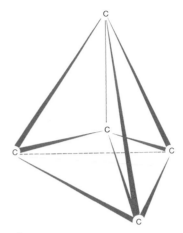

Figure 2

- **Diamonds are very hard.** Carbon atoms in diamond are linked by very strong covalent bonds. This makes diamond hard. Another reason for its hardness is that the atoms are not arranged in layers so they cannot slide over one another like the atoms in metals. Diamond is one of the hardest known substances. Most of its industrial uses depend on this hardness. Diamonds which are not good enough for gems are used in glass cutters and in diamond studded saws. Powdered diamonds are also used as abrasives for smoothing very hard materials.

- **Diamond has a very high melting point.** Carbon atoms in diamond are held in the crystal structure by very strong covalent bonds. This means that the atoms cannot vibrate fast enough to break away from their neighbours until very high temperatures are reached. So, the melting point of diamond is very high.

- **Diamond does not conduct electricity.** Diamond does *not* conduct electricity, unlike metals and graphite. In metals and graphite, some of the *outer* electrons are not strongly attached to any nucleus. They move towards the positive terminal when metals and graphite are connected to a battery. In diamond, however, the *outer* electrons of each carbon atom are held firmly in covalent bonds. So, diamond does not conduct electricity.

This diamond is being polished

Questions

1 Why is diamond called a giant molecule?

2 (a) Make a list of uses of diamond.
   (b) How do these uses depend on the properties and structure of diamond?

3 The largest natural diamond is the Cullinan diamond. This weighs about 600 g.
   (a) How many moles of carbon does it contain? (C = 12)
   (b) How many atoms of carbon does it contain?

4 Why are diamond cutters used to cut glass?

5 'Diamonds are a girl's best friend'. Is this true? What do you think?

4 Carbon: Graphite

Graphite is the second important allotrope of carbon. Figure 1 shows a model of part of the structure of graphite. Notice that the carbon atoms are arranged in parallel layers. Each layer contains millions and millions of carbon atoms arranged in hexagons. Each carbon atom is held strongly in its layer by covalent bonds, so every layer is a **giant molecule**. The distance between neighbouring carbon atoms in the same layer is only 0.14 nm, but the distance between the layers is 0.34 nm.

A can of lubricating oil containing graphite. Graphite has a layered structure. The layers slip and slide over one another very easily. This makes graphite an excellent material to improve the lubricating action of oils.

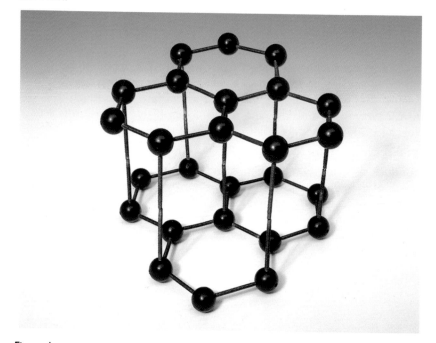

Figure 1
A model of the structure of graphite. Notice the layers of hexagons, one on top of the other

Properties and uses of graphite

● **Graphite is a lubricant.** In graphite, each carbon atom is linked by strong covalent bonds to three other atoms in its layer. But, the layers are 2½ times further apart than carbon atoms in the same layer. This means that the forces between the layers are weak. If you rub graphite, the layers slide over each other and onto your fingers. This property has led to the use of graphite as the 'lead' in pencils and as a lubricant. The layers of graphite slide over each other like a pile of wet microscope slides (figure 2). The wet slides stick together and it is difficult to pull them apart, but a force parallel to the slides pushes them over each other easily and smoothly.

● **Graphite has a very high melting point.** Although the layers of graphite move over each other easily, it is difficult to break the bonds between carbon atoms within one layer. Because of this, graphite does not melt until 3730° C and it does not boil until 4830° C. So, it is used to make crucibles for molten metals. The bonds between carbon atoms in the layers of graphite are so strong that graphite fibres with the layers arranged along the fibre are stronger than steel. These fibres are used to reinforce metals and broken bones.

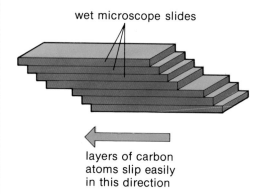

wet microscope slides

layers of carbon atoms slip easily in this direction

Figure 2
The layers in graphite slide over each other like wet microscope slides

• **Graphite conducts electricity.** The bonds *between* the layers of graphite are fairly weak. The electrons in these bonds move along the layers from one atom to the next when graphite is connected to a battery. So graphite will conduct electricity, unlike diamond and other non-metals. Because of this unusual property, graphite is used as electrodes in industry and as the positive terminals in dry cells.

Graphite fibres have been used to reinforce the shaft of this badminton racket

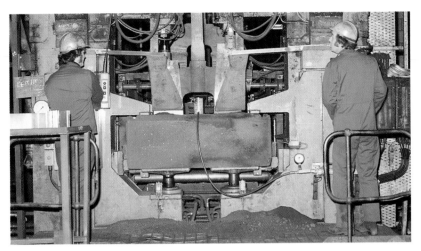

This large graphite anode is being prepared for use in the electrolytic extraction of aluminium

Charcoal

X-ray studies show that charcoal contains tiny crystals with a similar structure to graphite. The spaces between the layers of carbon atoms in finely powdered charcoal can trap other atoms and molecules. The biggest single use of powdered charcoal is in the sugar industry where it is used to absorb coloured impurities from brown sugar and syrup. Powdered charcoal can also absorb large volumes of many gases. One gram of charcoal will absorb 380 cm^3 of sulphur dioxide or 235 cm^3 of chlorine at room temperature, but oxygen is not readily absorbed. Because of this, charcoal is used in gas masks to protect the wearer from poisonous (toxic) gases.

Gas masks contain powdered charcoal which absorbs poisonous gases

5 Simple Molecular Substances

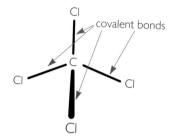

Figure 1
Tetrachloromethane (carbon tetrachloride) is a simple molecular substance. In tetrachloromethane, the carbon atom and four chlorine atoms are held together by strong covalent bonds

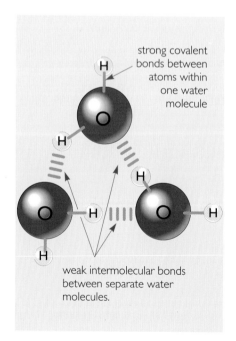

Figure 2
Covalent bonds and intermolecular bonds in water

Oxygen and water are good examples of simple molecular substances. They are made of simple molecules (each containing a few atoms). Their formulas and structures are shown near the top of table 1. Most other non-metals and non-metal compounds are also made of simple molecules. For example, hydrogen is H_2, chlorine is Cl_2, iodine is I_2, carbon dioxide is CO_2 and tetrachloromethane is CCl_4. Sugar ($C_{12}H_{22}O_{11}$) has much larger molecules than these substances, but it still counts as a simple molecule.

In these simple molecular substances, the atoms are held together in each molecule by strong covalent bonds (figure 1). The formulas and structures of a few simple molecular substances are shown in table 1.

Name	Molecular formula	Model of structure
Hydrogen	H_2	
Oxygen	O_2	
Water	H_2O	
Methane	CH_4	
Hydrogen chloride	HCl	
Iodine	I_2	
Carbon dioxide	CO_2	

Table 1: Formulas and structures of some simple molecular substances

Properties	Iodine	Sugar
Appearance	Soft dark grey crystals	Soft white crystals
Smell	Sharp disinfectant smell	Faint sweet smell
What happens on gentle heating?	Vaporises easily forming a purple vapour	Melts easily to a clear liquid
Does the solid conduct electricity?	No	No
Does the liquid conduct electricity?	No	No

Table 2: Properties of iodine and sugar

This butcher is using 'dry ice' (solid carbon dioxide) to keep meat cool and bacteria-free during mincing. After mincing, the 'dry ice', which is a simple molecular substance, will evaporate rapidly without spoiling the meat

Properties of simple molecular substances

Simple molecular substances have similar properties. These properties are shown by iodine (a non-metal element) and sugar (a non-metal compound) in table 2.

Look carefully at table 2. What properties do iodine and sugar have in common?

The properties of simple molecular substances can be explained in terms of their structure. The molecules in these substances have no electrical charge (like ions in ionic compounds). So there are no electrical forces holding them together. But as molecular substances do form liquids and solids, there must be some forces holding their molecules together. These weak forces between the separate molecules are called **intermolecular bonds** or **van der Waals' bonds** (figure 2).

● **Simple molecular substances are soft.** The separate molecules in simple molecular substances, such as I_2, are usually further apart than atoms in metal structures and further apart than ions in ionic structures. The forces between the molecules are only weak and the molecules are easy to separate. Because of this, crystals of these substances are usually soft.

● **Simple molecular substances have low melting points and boiling points.** It takes less energy to separate the molecules in simple molecular substances than to separate ions in ionic compounds, or atoms in metals. So, simple molecular compounds have lower melting points and lower boiling points than ionic compounds and metals.

● **Simple molecular substances do not conduct electricity.** Simple molecular substances have no mobile electrons like metals. They do not have any ions either. This means that they cannot conduct electricity as solids, as liquids or in aqueous solution.

Questions

1 Explain the following:
covalent bond; intermolecular bond; simple molecule.

2 List the main properties of simple molecular substances.

3 Simple molecular substances often have a smell, but metals do not. Why is this?

4 A substance is a poor conductor of electricity in the solid state. It melts at 217° C and boils at 685° C. Could this substance be (i) a metal, (ii) a non-metal, (iii) a giant molecule, (iv) an ionic solid, (v) a simple molecular solid?

5 What properties does butter have that show that it contains simple molecular substances?

6 Molecular Compounds

If the nucleus of an atom was magnified a million, million times, it would be big as a pea and the total volume of the atom in which the electrons move would be as large as Westminster Abbey

Metals can be mixed to form alloys, but they *never* react with each other to form compounds. For example, zinc and copper will form the alloy brass, but the two metals cannot react chemically because they both want to lose electrons and form positive ions.

Unlike metals, two non-metals can react with each other and form a compound even though they both want to gain electrons. These *non-metal compounds* are composed of simple molecules as we saw in the last unit. They are therefore called **molecular compounds** and include water (H_2O), carbon dioxide (CO_2), sugar ($C_{12}H_{22}O_{11}$) and ammonia (NH_3).

Forming molecular compounds—electron sharing

All atoms have a small positive centre called a **nucleus**, surrounded by a larger region of negative charge. The negative charge consists of electrons. It is balanced by positive charge in the nucleus, so that the whole atom is neutral. Almost all the mass of the atom is concentrated in the nucleus (figure 1). Different atoms have different numbers of electrons. Hydrogen atoms are the smallest with only one electron, helium atoms have two electrons and oxygen atoms have eight electrons.

When two non-metals react to form a molecule, the regions of electrons in their atoms overlap so that each atom gains negative charge. The positive nuclei of both atoms attract the electrons in the region of overlap and his holds the atoms together (figure 2). This type of bond formed by *electron sharing* between non-metals is a **covalent bond**. Notice that covalent bonding, like ionic bonding, involves attraction between opposite charges. Covalent bonds hold the atoms together *within* a molecule but there are also **intermolecular bonds** holding the separate molecules together in molecular liquids like water and molecular solids like sugar. For example, in water there are strong covalent bonds between the two hydrogen atoms and the oxygen atom *within* each molecule of H_2O and also weak intermolecular bonds between the different molecules of H_2O.

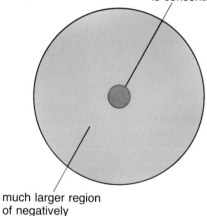

very small positive nucleus where the mass is concentrated

much larger region of negatively charged electrons

Figure 1
A simple picture of atomic structure

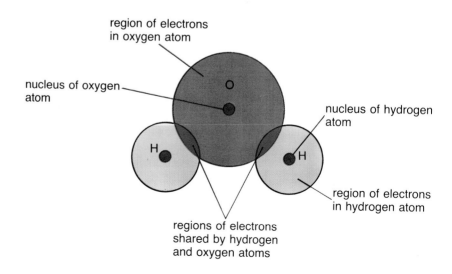

region of electrons in oxygen atom

nucleus of oxygen atom

nucleus of hydrogen atom

region of electrons in hydrogen atom

regions of electrons shared by hydrogen and oxygen atoms

Figure 2
The simple structure of a molecule of water

Formulas of molecular compounds

The table below shows the formulas and structures of some well-known molecular compounds. The structures are written so that the number of covalent bonds (drawn as a line —) to each atom can be seen. Notice that hydrogen can form 1 bond with other atoms (H—), so its **combining power** or **valency** is 1. The combining powers of chlorine and bromine are also 1. Oxygen and sulphur both form 2 bonds with other atoms (—O— and —S—). Their combining power is therefore 2. Nitrogen atoms form 3 bonds and carbon atoms form 4 bonds, so their valencies are 3 and 4 respectively.

Although we can predict the formulas of molecular compounds from the number of bonds which the atoms form, the only sure way of knowing a formula is by chemical analysis. For example, carbon forms 4 bonds and oxygen forms 2 bonds, so we would predict that carbon and oxygen will form a compound O=C=O. (Each bond is represented by a single line, so two lines show that there is a double bond between the atoms.) This compound, carbon dioxide, does exist, but so does carbon monoxide, CO, which we would not predict.

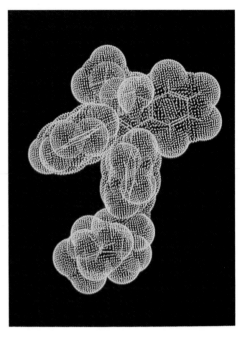

This model of a molecule was drawn using a computer to analyse the results obtained during the investigation of a complex organic substance

Compound	Formula	Structure
Hydrogen chloride	HCl	H—Cl
Hydrogen bromide	HBr	H—Br
Hydrogen iodide	HI	H—I
Water (hydrogen oxide)	H_2O	H—O—H
Hydrogen sulphide	H_2S	H—S—H
Ammonia (hydrogen nitride)	NH_3	H, H N, H (structure)
Methane	CH_4	H, H—C—H, H (structure)
Carbon dioxide	CO_2	O=C=O
Carbon disulphide	CS_2	S=C=S
Tetrachloromethane (carbon tetrachloride)	CCl_4	Cl, Cl—C—Cl, Cl (structure)

The formulas and structures of some well known molecular compounds

Questions

1 Explain the following:
non-metal compound; molecular compound; covalent bond; nucleus; intermolecular bond.

2 Assuming the usual combining powers of the elements, draw the structures of the following compounds: (Show each bond as a line —.)
sulphur dichloride; dichlorine oxide; tetrabromomethane (carbon tetrabromide); nitrogen triiodide; hydrogen peroxide, H_2O_2; ethane, C_2H_6; ethene (C_2H_4).

7 Carbon Dioxide

Carbon dioxide is an important simple molecular compound. It links respiration and photosynthesis and is produced when carbon compounds burn. Carbon dioxide also has some important uses.

When a bottle of soda water is opened, the liquid fizzes because the pressure falls and gas escapes

A fireman demonstrates the use of a small carbon dioxide fire extinguisher

- **Soda water and fizzy drinks.** Solutions of carbon dioxide in water have a pleasant taste—the taste of soda water. Soda water and other fizzy drinks are made by dissolving carbon dioxide in them at high pressure. When a bottle of the drink is opened, it fizzes because the pressure falls and carbon dioxide gas can escape from the liquid.

- **Fire extinguishers.** Liquid and gaseous carbon dioxide at high pressure are used in fire extinguishers. When the extinguisher is used, carbon dioxide pours out and smothers the fire. Carbon dioxide is heavier than air so it covers the fire and stops oxygen getting to it. The fire 'goes out' because carbon dioxide does not burn and substances will not burn in it.

- **Refrigeration.** Solid carbon dioxide is used for refrigerating ice-cream, soft fruit and meat. The solid carbon dioxide is called 'dry ice' because it resembles ice. It is colder than ordinary ice and sublimes without going through the messy liquid stage. This is why it is called 'dry ice' or 'Dricold'.

Making carbon dioxide

Strong acids, like hydrochloric acid, sulphuric acid and nitric acid, react with carbonates to form carbon dioxide and water.

$$2H^+(aq) + CO_3^{2-}(s) \rightarrow H_2O(l) + CO_2(g)$$
$$\text{acid} \qquad \text{carbonate}$$

Small amounts of carbon dioxide are usually prepared from marble chips (calcium carbonate) and dilute hydrochloric acid (figure 1).

$$CaCO_3(s) + 2HCl(aq) \rightarrow CaCl_2(aq) + H_2O(l) + CO_2(g)$$

The carbon dioxide may be collected by downward delivery or over water.

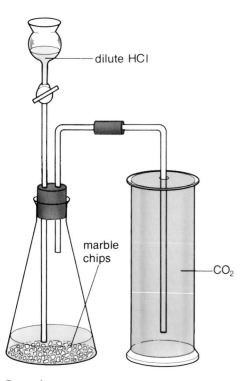

dilute HCl

marble chips

CO₂

Figure 1
Making CO₂ in the laboratory. **Wear eye protection** if you are making CO₂ in this way

Properties of carbon dioxide

Carbon dioxide is a typical non-metal oxide. It is acidic, gaseous and simple molecular. Figure 2 shows some other properties of carbon dioxide. Notice that it is slightly soluble in water. The dissolved gas provides water plants, like seaweed, with the carbon dioxide they need for photosynthesis. About 1% of the gas which dissolves in water reacts to form carbonic acid.

$$H_2O + CO_2 \rightarrow H_2CO_3$$

The solution of carbonic acid is a very weak acid. It turns blue litmus paper only a purplish-red.

Testing for carbon dioxide with limewater

The test for carbon dioxide uses its acidic property. Lime water is calcium hydroxide solution—a dilute alkali. When carbon dioxide is bubbled into lime water, the liquid goes milky with a white precipitate of calcium carbonate. Why does this precipitate form? First, the carbon dioxide reacts with OH^- ions in the alkali to form carbonate.

$$CO_2(g) + 2OH^-(aq) \rightarrow CO_3^{2-}(aq) + H_2O(l)$$

Then, CO_3^{2-} ions react with calcium ions in the lime water to form insoluble calcium carbonate.

$$Ca^{2+}(aq) + CO_3^{2-}(aq) \rightarrow CaCO_3(s)$$

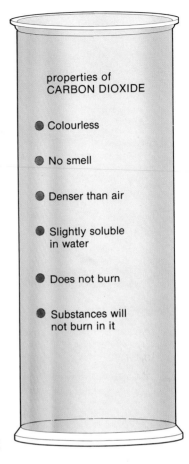

properties of
CARBON DIOXIDE

● Colourless

● No smell

● Denser than air

● Slightly soluble in water

● Does not burn

● Substances will not burn in it

Figure 2

These workmen are using 'dry ice' to shrink-fit an axle into a huge cog wheel. Why does the axle shrink when it is surrounded by solid CO_2?

Questions

1 List the important uses of carbon dioxide. For each use, explain why carbon dioxide is used.
2 How does carbon dioxide link respiration and photosynthesis?
3 Give two reasons why 'dry ice' is better than ordinary ice for refrigeration.
4 Carbon dioxide can be poured from a gas jar onto a lighted candle and the candle goes out. What properties does this simple experiment show for carbon dioxide?
5 (a) How is carbon dioxide obtained from a carbonate?
  (b) Write an equation for the reaction in part (a).
  (c) How would you show that limestone is a carbonate?

8 Ionic Compounds

Forming ionic compounds—electron transfer

Ionic compounds form when metals react with non-metals. For example, when sodium burns in chlorine, sodium chloride is formed. This contains sodium ions (Na^+) and chloride ions (Cl^-).

$$Na + Cl \rightarrow Na^+ \quad Cl^-$$

sodium atom   chlorine atom   sodium ion   chloride ion

These ions form by *transfer of electrons*. During the reaction, each sodium atom gives up one electron and forms a sodium ion:

$$Na \rightarrow Na^+ + e^-$$

The electron is taken by a chlorine atom to form a chloride ion:

$$Cl + e^- \rightarrow Cl^-$$

When ionic compounds form, metal atoms *lose* electrons and form *positive* ions, whereas non-metal atoms *gain* electrons and form *negative* ions. This transfer of electrons from metals to non-metals explains the formation of ionic compounds. Figure 1 shows what happens when calcium reacts with oxygen to form calcium oxide. In this case, two electrons are transferred from each calcium atom to each oxygen atom.

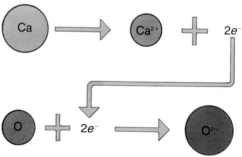

calcium atom   oxygen atom   calcium ion   oxide ion

calcium oxide

Figure 1

Chalk cliffs are composed of an ionic compound containing calcium ions (Ca^{2+}) and carbonate ions (CO_3^{2-})

Bonding and properties of ionic compounds

In solid ionic compounds, the ions are held together by the attraction between positive ions and negative ions. Figure 2 shows how the ions are arranged in sodium chloride. Notice that Na^+ ions are surrounded by Cl^- ions and that the Cl^- ions are surrounded by Na^+ ions.

This kind of arrangement in which a large number of atoms or ions are packed together in a regular pattern is called a **giant structure**. The force of attraction between oppositely-charged ions is called an **ionic or electrovalent bond**. The strong ionic bonds hold the ions together very firmly. This explains why ionic compounds:

Figure 2
The arrangement of ions in one layer of a sodium chloride crystal

1 are solids at room temperature with high melting points;

2 are hard substances;

3 conduct electricity when molten or aqueous;

4 cannot conduct when solid because the ions cannot move freely.

Name of salt	Formula
calcium nitrate	$Ca^{2+}(NO_3^-)_2$ or $Ca(NO_3)_2$
zinc sulphate	$Zn^{2+}SO_4^{2-}$ or $ZnSO_4$
iron(III) chloride	$Fe^{3+}(Cl^-)_3$ or $FeCl_3$
copper(II) bromide	$Cu^{2+}(Br^-)_2$ or $CuBr_2$
sodium carbonate	$(Na^+)_2CO_3^{2-}$ or Na_2CO_3
potassium iodide	K^+I^- or KI

Table 1: The names and formulas of some salts

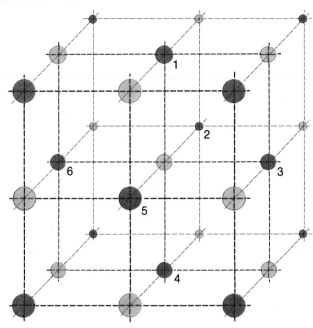

Figure 3
A model of the structure of sodium chloride. The large green balls represent Cl^- ions ($A_r = 35.5$). The smaller silver balls represent Na^+ ions ($A_r = 23.0$)

Formulas of ionic compounds—balancing charges

The formulas of ionic compounds, like sodium chloride (NaCl) and calcium oxide (CaO), could be obtained by balancing the charges on the positive ions with those on the negative ions. For example, the formula of calcium chloride is $Ca^{2+}(Cl^-)_2$ or simply $CaCl_2$. Here, two Cl^- ions balance the charge on one Ca^{2+} ion. The formula $CaCl_2$ has a small 2 after the Cl to show that 2 Cl^- ions are needed for one Ca. These formulas show the ratio of the numbers of ions present in the ionic compound.

Can you see that the number of charges on an ion is a measure of its **combining power** or **valency**? Na^+ has a combining power of 1, whereas Ca^{2+} has a combining power of 2. Na^+ can combine with only one Cl^- to form Na^+Cl^-, whereas Ca^{2+} can combine with two Cl^- ions to form $Ca^{2+}(Cl^-)_2$.

Elements such as iron, which have two different ions (Fe^{2+} and Fe^{3+}), have two valencies. Thus iron can form two different compounds with chlorine—iron(II) chloride, $FeCl_2$, and iron(III) chloride, $FeCl_3$.

Table 1 shows the names and formulas of some salts. Notice that the formula of calcium nitrate is $Ca(NO_3)_2$. The brackets around NO_3^- show that it is a single unit containing one nitrogen and three oxygen atoms with one negative charge. Thus, two NO_3^- ions balance one Ca^{2+} ion. Other ions like SO_4^{2-}, CO_3^{2-} and OH^- must also be regarded as single units and put in brackets when there are 2 or 3 of them in a formula.

Questions

1 Which of the following substances conduct electricity (i) when liquid; (ii) when solid:
diamond; potassium chloride; copper; carbon disulphide; sulphur?

2 Look carefully at figures 2 and 3.
  (a) How many Cl^- ions surround one Na^+ ion in the three dimensional crystal?
  (b) How many Na^+ ions surround one Cl^- ion in the three dimensional crystal?

3 Write the symbols for the ions and the formulas of the following compounds:
potassium hydroxide; iron(III) nitrate; barium chloride; sodium carbonate; silver sulphate; calcium hydrogencarbonate; aluminium oxide; zinc bromide; copper(II) nitrate; magnesium sulphide.

4 Sodium fluoride and magnesium oxide have the same crystal structure and similar distances between ions. The melting point of NaF is 992° C, but that of MgO is 2640° C. Why is there such a big difference in their melting points?

9 Salt—An Important Ionic Compound

Tennis players and people who work in hot places usually take salt tablets to replace the salt they lose by sweating. Sweat is mainly salt solution

Ionic compounds are present in the sea and in the Earth's crust. Many rocks contain ionic compounds. These include rock salt (Na^+Cl^-), limestone ($Ca^{2+}CO_3^{2-}$) and iron ore ((Fe^{3+})$_2$(O^{2-})$_3$). Clay, sandstone and granite also contain ionic compounds.

Salt in our diet

Salt (sodium chloride) is one of the most important ionic compounds. It is an essential mineral in our diet. Most foods contain salt but some foods are saltier than others. Our diet must contain the right amount of salt. Too much salt may cause high blood pressure. Too little salt causes sharp pains ('cramp') in our muscles.

Cattle licking a salt block. The salt block contains various salts to supplement the diet of the cattle

Properties of ionic compounds

Ionic compounds
- *are hard*
- *have high melting points*
- *have high boiling points*
- *do not conduct when solid*
- *conduct when liquid and in aqueous solution*

The box above shows the important properties of ionic compounds. These properties can be explained in terms of their giant structure of ions.

Large amounts of impure sodium chloride (crushed rock salt) are used for gritting and de-icing roads. The salt mixes with the ice and lowers its melting point. Mixtures of ice and salt will melt at temperatures down to $-22°$ C. This means that the temperatures can be well below $0°$ C and the roads still do not ice up

- **Hardness.** In order to cut a lump of sodium chloride, the ions must be separated. This is very difficult because each ion is held in the crystal lattice by strong attractions from the ions of opposite charge around it.

- **Melting and boiling.** In ionic solids, the ions vibrate about fixed positions. As the temperature rises, the ions vibrate more and more.

Eventually, the ions vibrate so much that they 'escape' from their places in the crystal and slide freely around each other. When this happens, the solid is melting. The forces between ions of opposite charge are so strong that ionic compounds have high melting points and even higher boiling points.

Some ionic compounds, such as magnesium oxide and aluminium oxide, have *extremely* high melting points. Because of this, they can be used as a lining for furnaces. They are called **refractory materials**.

● **Conductivity.** Solid ionic compounds cannot conduct electricity. The ions are held in the crystal and cannot move towards the electrodes. When the solid melts, and when it is dissolved in water, the ions are free to move. So molten and aqueous ionic compounds *will* conduct.

Salt—an important source of other substances

The electrolysis of *molten* sodium chloride is used to produce sodium and chlorine (section D, unit 4). The electrolysis of saturated sodium chloride *solution* (brine) is used to manufacture chlorine, hydrogen and sodium hydroxide (figure 1).

During electrolysis, chloride ions in the brine are attracted to the graphite anodes. At the anodes, they are converted into chlorine gas.

$$\textbf{Anode (+)} \qquad 2Cl^-(aq) \rightarrow 2e^- + Cl_2(g)$$

At the same time, sodium ions in the brine are attracted to the flowing mercury cathode. The sodium ions take electrons from the cathode and form sodium metal. This dissolves in the mercury to form an amalgam (alloy).

$$\textbf{Cathode (−)} \qquad 2Na^+(aq) + 2e^- \rightarrow 2Na(s)$$

The sodium/mercury amalgam flows out of the cell and into a second tank. Here, it reacts with water to produce hydrogen and sodium hydroxide.

$$2Na/Hg(l) + 2H_2O(l) \rightarrow 2NaOH(aq) + H_2(g) + 2Hg(l)$$

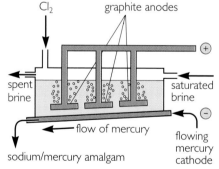

Figure 1
The electrolysis of saturated sodium chloride solution (brine)

Questions

1 Draw a diagram or design a poster to illustrate the uses of sodium chloride (salt) and the materials which are obtained from it.

2 Solid sodium chloride does not conduct electricity, but liquid sodium chloride conducts well.
   (a) Explain this statement.
   (b) Write equations for the processes at the electrodes when liquid NaCl conducts.

3 Give three important uses of sodium chloride as pure salt, rock salt or brine. Explain why sodium chloride has these uses.

4 Substance X melts at a high temperature. Liquid X conducts electricity.
   (a) Which of the following could be X?
   calcium chloride, starch, copper, sulphur, polythene, bronze, carbon disulphide, zinc oxide.
   (b) Explain your answers to part (a).

5 Explain the following points.
   (a) Cattle require a salt lick during hot summer months.
   (b) The use of crushed rock salt on icy roads does have a drawback for motorists.
   (c) Salt is very unreactive, but sodium and chlorine, from which it is made, are both very reactive.

10 From Sand to Glass

Different kinds of glass

We use different kinds of glass for different purposes. Some of these are illustrated in the photos below.

Bottles and windows are made from cheap soda lime glass

Glass dishes and ornaments use lead glass which is harder and shinier

Laboratory glassware and glass ovenware are made from borosilicate glass which is heat resistant

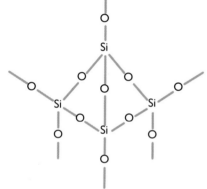

Figure 1
The structure of pure silicon dioxide glass. (For simplicity, the structure is not drawn in three dimensions)

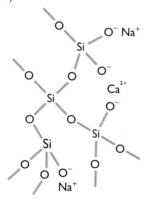

Figure 2
The structure of soda-lime glass. (For simplicity, the structure is not drawn in three dimensions)

Making glass

Glass is usually made by heating a mixture of metal oxides and metal carbonates with pure sand (silicon dioxide, SiO_2) in a furnace. At the high temperatures involved, any carbonates decompose to oxides and bubbles of carbon dioxide form in the mixture. The glass is heated to make sure that all the bubbles escape and this produces a runny, molten liquid. The liquid is cooled until it is thick enough to be moulded or blown into different shapes. It is then allowed to cool further until it sets solid.

The structure of glass

Glass can be made from pure silicon dioxide. This melts at 1700° C to give a thick, viscous liquid. On cooling, the liquid forms a glassy, transparent solid. This glass has a giant structure of atoms. Every silicon atom is bonded to four oxygen atoms (figure 1).

This three-dimensional structure with strong covalent bonds between silicon and oxygen atoms makes the glass very hard. Unfortunately, the high melting point of this glass makes it expensive to manufacture and difficult to mould. Because of this, most glasses are made from a mixture of silicon dioxide with metal oxides. The metal oxides lower the melting point. For example, soda lime glass is made from sand (silicon dioxide), limestone (calcium carbonate) and sodium carbonate. The carbonates decompose to oxides on heating. So the glass has a giant structure of atoms *and* ions (figure 2). It has both ionic and covalent bonding.

The properties of glass

Glass is
- hard
- chemically unreactive
- easy to clean
- heat resistant
- an insulator
- transparent
- fairly cheap
- easily moulded

Recycling glass

More than six thousand million glass bottles and jars are sold in Britain each year. Most of these are thrown away. A small proportion (perhaps one tenth) are recycled. This is disappointing for a number of reasons.

- Glass containers are relatively easy to clean, sterilize and refill.

- Glass can easily be crushed and added to furnaces to make new containers.

- Glass is made from materials such as sand, limestone and salt which have to be extracted from the earth by processes which spoil the landscape and use up our energy supplies.

- Re-using and recycling glass could save huge amounts of energy.

Glass is sometimes used to encase nuclear waste before it is transported. Why do you think glass is used for this purpose?

Questions

1 Get into a group with 2 or 3 others. Look at the properties and the structure of glass described in this unit.

  (a) In your group, discuss how the structure of glass results in its properties.

  (b) Why do you think glass cracks and breaks if it is dropped or if the temperature suddenly changes?

2 Glass insulators are often used on electricity pylons. Make a list of the properties of glass which make it particularly useful for this purpose.

3 Look at the energy costs involved in recycling glass, in table 1.

  (a) What is the total energy cost of collecting, processing and delivering enough waste glass to the furnace to produce one kilogram of recycled glass?

Recycling process	Energy cost per kg of recycled product/kJ
Collecting waste glass	350
Processing waste glass	100
Delivering waste glass to glass furnace	50

Table 1

  (b) The energy cost of extracting, processing and delivering enough raw materials to the furnace to produce one kilogram of *new* glass, is 4500 kJ. How much energy is saved, per kilogram, if glass is recycled?

  (c) How much energy is saved, per *tonne*, if glass is recycled? (1 tonne = 1000 kg)

  (d) Suppose that all the fuels used to provide this energy (petrol, diesel, fuel oil) produce 40 000 kJ per litre. How many litres of fuel are saved, per tonne, if glass is recycled?

  (e) Suppose that all the fuels used in glass manufacture cost 50p per litre. How much money is saved, per tonne, if glass is recycled?

4 (a) Why do you think so few glass containers are re-used at present?

  (b) What proportion of glass containers do you think could be re-used or recycled? Explain your answer.

  (c) How could we encourage people to re-use or recycle glass?

11 From Clay to Ceramics

What are ceramics?

The word 'ceramic' comes from a Greek word meaning 'pottery' or 'burnt clay'. Clay has been dug from the ground and heated to make pots, bricks and ornaments since prehistoric times.

This piece of pottery is being made using a machine

This photo shows where china clay is extracted. China clay is the best quality clay for pots and ornaments

Finished pieces of pottery after glazing and firing

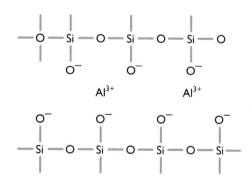

Figure 1
A simplified diagram to show the bonding in clay

What is clay?

Clay is a mixture of several materials. The best quality clay (china clay), which is used to make crockery and ornaments, contains a high proportion of kaolinite. Kaolinite contains silicon, oxygen and aluminium atoms. The silicon and oxygen atoms are joined by strong covalent bonds and there are ionic bonds between the oxygen and aluminium. These atoms are linked together in separate flat layers in a giant structure. These flat layers are described as **two-dimensional structures**.

Figure 1 gives a very simplified picture of the bonding and structure in raw clay.

Notice the similarity between the structure of raw clay and those of sand and glass. There is, however, one important difference. Clay has a flat, two-dimensional structure whereas the structures of sand and glass are three-dimensional.

Moulding and firing clay

When clay is wet, water molecules get between the layers and allow them to slide over one another. This is why wet clay is soft and slippery and can easily be moulded into different shapes. The water acts as a lubricant. When clay is left to dry, most of the water between the layers evaporates. The clay also shrinks and the layers get closer so that they cannot move over each other so easily.

When clay is heated in a furnace at about 1000° C (**fired**), it forms a ceramic. This is hard, gritty and rigid. During firing, all the water molecules are driven out of the clay and complicated chemical changes take place. Atoms in one layer form bonds with those in the layers above and below. So the flat, two-dimensional structure in clay changes to a cross-linked, three-dimensional structure in the ceramic. The structure of ceramics is therefore very similar to that of glass.

Using ceramics

Ceramics and glasses are very important in our homes and industries. They have high melting points and are chemically unreactive. They contain elements like silicon and aluminium already combined with oxygen. So they do not burn when heated in air. In this respect, they differ from metals and plastics.

Because of their heat resistance, ceramics are used to line most industrial furnaces. Any other material would either melt or burn. They are used in furnaces for the manufacture of iron, steel, glass and cement. The bricks used for these furnaces are made from fireclay. This contains a higher proportion of aluminium than other clays and it produces bricks which are more heat resistant and chemically inert.

Probably the most important use of ceramics is for crockery. At one time, crockery was shaped by hand on a potter's wheel, but now most of it is shaped and moulded by machines, as shown in the photo on the opposite page.

Hollow articles like teapots and jugs are made by pouring runny clay into moulds. The moulds are made of porous plaster. This absorbs water from the clay and a solid lining forms on the inside of the mould. When this lining is sufficiently thick, the rest of the runny clay is poured out. The article can then be removed from the mould and dried.

Bricks and tiles are made by pressing dryish clay into steel moulds. Once the clay has been shaped or moulded, it must be dried before firing. Otherwise, the water will evaporate and shatter the clay during firing.

Apart from bricks and roof tiles, most ceramic articles are covered with a **glaze**. The glaze gives the ceramic a smooth, glassy, waterproof surface. After the first firing, the pottery is dipped in a glaze and then fired again. The glaze consists of finely powdered oxides of silicon and metals suspended in water. When the glazed pottery is fired a second time, the glaze is converted to glass.

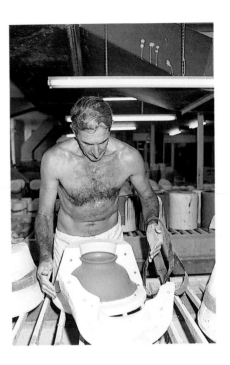

A dried clay article being removed from its mould before firing

Questions

1 Get into a group with 2 or 3 others. Discuss the following points and try to explain them.
  (a) Clay is soft and slippery, but ceramic is hard and gritty.
  (b) Dry, crumbly clay can be made soft and easily moulded by adding water.
  (c) Ceramic cannot be converted back to clay by adding water to it.
2 (a) Make a list of the similarities between glasses and ceramics.
  (b) Make a list of the differences between glasses and ceramics.
  (Hint: You might like to look at the properties of glass described in the last unit.)
3 (a) Draw a flow diagram to show the stages in making a teapot from china clay.
  (b) Why are the moulds used for teapots made of porous material?
  (c) Why must clay be dried before it is fired?
  (d) What is the advantage of coating articles like teapots with a glaze?

Section H: Activities

1 From bamboo to carbon fibre

The table below shows the changes in the world pole vault record since 1920.

Year	1920	1930	1940	1950	1960	1970	1980	1990
Height/m	4.2	4.3	4.5	4.6	4.8	5.3	5.7	6.1

1 Plot a graph of the world pole vault record against the year.
2 When did the record improve by the greatest amount?

The large increase in the pole vault record which you found in answer 2 resulted from the use of glass fibre poles. Glass fibre is known as a **composite**, made up of two different materials. It is made of fibres of glass embedded in plastic. The resulting composite has the strength of one material and the flexibility of the other. Before 1960, poles were made of aluminium. Before aluminium, they were made of bamboo. Recently, poles have been made of carbon fibre and this has proved even better than glass fibre.

3 What is a 'composite'?
4 Which material in glass fibre gives it (i) strength, (ii) flexibility?
5 Why do you think glass fibre was better than aluminium for poles?
6 Why do you think aluminium was better than bamboo?
7 Why is carbon fibre particularly good for poles?

Pole vaulting during the 1920s. What is the pole made from?

2 Explaining the structure of rubber

Natural rubber is a hydrocarbon. It contains long chains of carbon atoms with hydrogen atoms attached to them (figure 1). Each rubber molecule contains between 10 000 and 50 000 carbon atoms. Notice that every fourth bond between the carbon atoms is a double bond.

In natural rubber, millions of these long chains are tangled together like tangled pieces of wire.

Unfortunately, natural rubber is only partly elastic. When it is pulled or deformed the molecules move over one another and do not return to their original position when the force is removed (figure 2). Pulling a tangle of wires will have the same effect.

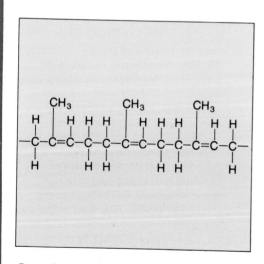

Figure 1
Part of a molecule of natural rubber

tangled chains in unstretched rubber

chains in stretched rubber

Figure 2
Stretching natural rubber

The structure of the natural rubber must be changed slightly to make it more elastic. If we use the wire model again, it is fairly obvious that the sliding and slipping can be prevented if pairs of wires are fixed together where they cross. But, how can this be done with natural rubber?

The answer lies in the carbon–carbon double bonds. If sulphur is heated with natural rubber, the carbon–carbon double bonds break open and sulphur atoms form links between the long carbon chains (figure 3). This process is called vulcanization.

The links between the rubber molecules stop the chains sliding over one another too much, yet the rubber stays elastic.

1 Make a model of the structure of rubber using four or five tangled pieces of wire.
2 Does your model behave like natural rubber when it is stretched?
3 Fix pairs of wire together using paper clips or short pieces of wire. How does this affect the stretching and elastic properties of your wire model?
4 How do you think the elasticity of rubber will change as the amount of sulphur used for vulcanization increases?
5 If possible, use ball and spoke molecular models to build a model of part of a molecule of natural rubber like that shown in figure 1.
6 Use your molecular model from part 5 to explore the way in which sulphur atoms can form crosslinks between the rubber molecules.

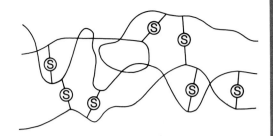

Figure 3
Sulphur links between molecules of natural rubber

Vulcanized rubber tyres coming off the production line

3 | Using X-rays to study materials

Most of our ideas about the structure of materials has come from X-ray crystallography. One of the most important pieces of X-ray crystallography involved finding the structure of penicillin. During the Second World War increased quantities of penicillin were needed for the treatment of wounds incurred by our fighting troops.

Scientists were asked to find ways of producing larger quantities of penicillin synthetically. But, before they could do this, they needed to know its structure. Dorothy Hodgkin used X-ray crystallography to find the structure of penicillin. In 1964, she received the Nobel Prize for her work on the structure of penicillin and other biological substances.

1 Find out what you can about the use of X-rays to study the structure of materials. (You could start by reading unit 1 in section H of this book.)
2 Design and prepare one page of notes, diagrams and photos, etc. to help you understand how X-rays are used to study the structure of materials.
3 Use a word processor to prepare a neat version of your notes. Try to use all the facilities of your word processor including different sized letters, diagrams, shading, brushwork, etc.
4 Print a copy of your notes. Add in any photos and diagrams that are necessary and then take a photocopy of the final version.

Dorothy Hodgkin

Section H: Study Questions

1 Titanium is the seventh most abundant element in the Earth's crust. One form in which it occurs is *rutile*, TiO_2. In extracting titanium from its ore, rutile is first converted to titanium(IV) chloride, $TiCl_4$, and this is then reduced to the metal by heating it with sodium or magnesium in an atmosphere of argon. Titanium(IV) chloride is a simple molecular covalent substance.

(a) Given that the titanium atom has four electrons used for bonding, draw a diagram to show the bonding in titanium(IV) chloride. (Only the outer electrons of the chlorine atoms should be shown.)

(b) Write a balanced equation for the reaction of titanium(IV) chloride with sodium.

(c) (i) In which physical state would you expect to find titanium(IV) chloride at room temperature?

(ii) Explain why the state of titanium(IV) chloride differs from that of sodium chloride at room temperature.

(d) Suggest a reason why it is necessary to carry out the reaction of titanium(IV) chloride with sodium in an atmosphere of argon.

(e) Titanium is expensive in spite of the fact that it is relatively abundant in the Earth's crust. Suggest a reason for this.

(f) Titanium is used in the structures of supersonic aircraft. Suggest *two* properties it might have that make it more suitable than other metals for this purpose.

(g) Titanium is a transition metal. State *two* properties, different from those in (f) which you would expect it to have.

ULEAC

2
port and docks 10 km

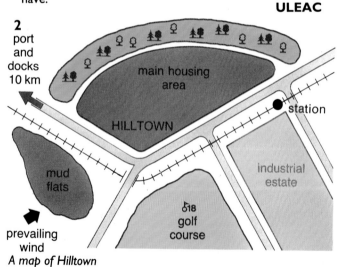

A map of Hilltown

A chemical company (ChlorChem) wants to build a new factory near Hilltown in order to manufacture chlorine and sodium hydroxide using brine.

(a) What problems does this pose?

(b) Where would you site this new factory?

(c) What are the reasons for your choice of site?

(d) The Council at Hilltown wants ChlorChem to build their factory on the mud flats. They have offered ChlorChem financial help if the factory is sited on the mud flats. What disadvantage will this site have for (i) ChlorChem, (ii) Hilltown?

3
Carborundum was one of the first abrasives to be made. Before the discovery of carborundum, powdered diamond was the most common abrasive.

When the first carborundum factory was opened, carborundum cost £4.40 per kilogram. After a few years, its price was 7p per kilogram.

Carborundum is a compound of carbon and silicon. It has a structure like diamond. Its chemical name is silicon carbide, SiC. Carborundum is manufactured by heating coke (carbon) with sand (silicon(IV) oxide) at 2500° C in a furnace.

1 What is an abrasive?

2 What are abrasives used for?

3 Suggest three important properties of an abrasive.

4 Why do you think carborundum replaced diamond as the most widely used abrasive?

5 Write a word equation and a balanced symbolic equation for the reaction of coke with sand to form carborundum.

6 Draw a diagram to show the structure of carborundum.

7 It has been suggested that carborundum made the Industrial Revolution possible during the nineteenth century. Why is this?

4 In order to find the formula of a compound of tin and chlorine, a stream of dry chlorine was passed over 3.0 g of tin in the apparatus below. When the reaction was complete, the liquid tin chloride was distilled into the receiver.

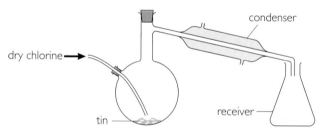

(a) Why is a condenser used?

(b) Make a neat sketch of the diagram.

(i) Show where the water should enter and leave the condenser.

(ii) The thermometer has been left out. Draw it in your sketch in the right place.

(c) Why should this experiment be done in a fume cupboard?

(d) Write down the weighings you would make to find the mass of tin chloride collected.

(e) In the experiment, 6.5 g of tin chloride should form.

(i) What mass of chlorine reacts with 3.0 g of tin?

(ii) How many moles of chlorine atoms ($Cl = 35$) combine with 3.0 g of tin?

(iii) How many moles of tin ($Sn = 120$) are there in 3.0 g?

(iv) How many moles of chlorine atoms combine with 1 mole of tin?

(v) What is the formula of the tin chloride?

(f) How would you check the purity of the tin chloride by a simple method?

SECTION I
Energy and Fuels

A summary of our energy sources. Look carefully at this picture. What examples of energy sources can you see?

1 Energy in Everyday Life

Everyone needs energy for living. We use energy from electricity or from burning fuels to keep us warm at home and at school. The African bush people in this photo warm themselves with the energy from an open wood fire

Everyday we use energy in our homes, schools and industries in thousands of different ways. We need energy to warm a room, cook a meal or light a torch. Although we use a lot of energy in our homes, industry uses far more. Energy is needed to turn raw materials like clay into useful things like bricks. It is also needed to mine coal and other minerals and to generate electricity. In fact,

> *Energy is essential to our society.*

Transferring energy to or from chemicals is the basis of chemical changes. Some chemical changes, such as the decomposition of limestone to lime, need heat to make them happen.

$$CaCO_3 \xrightarrow{\text{heat}} CaO + CO_2$$
$$\text{limestone} \qquad \text{lime} \quad \text{carbon dioxide}$$

Chemical reactions, which *take in* heat, are described as **endothermic**. Other chemical reactions *give out* heat. In these reactions, chemical energy is converted into heat. This happens when fuels such as coal, oil and natural gas are burnt.

$$CH_4 + 2O_2 \rightarrow CO_2 + 2H_2O + \text{heat}$$
$$\text{natural gas}$$

Chemical reactions like this, which *give out* heat, are described as **exothermic**.

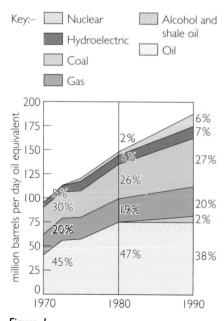

Key:–
- Nuclear
- Hydroelectric
- Coal
- Gas
- Alcohol and shale oil
- Oil

Figure 1
The percentage contributions that different fuels have made to the world's energy requirements since 1970

Most coal in Britain is used to generate electricity. This is why most power stations have huge coal heaps nearby. This photograph shows the coal-fired power station at Didcot in Oxfordshire with its huge supply of coal in the foreground

Everyday, the world needs energy equivalent to about 180 million barrels of oil. This energy is provided by a range of fuels. Figure 1 shows how the world's energy requirements have been provided by different fuels since 1970. Today, about two-fifths of our requirements are provided by oil, one fifth by gas and one quarter by coal. The remaining 15% of our energy requirements is provided by small amounts of alcohol, shale oil, nuclear power and hydroelectricity.

Foods such as fats and carbohydrates are biological fuels. They are important stores of energy. When foods are broken down (**metabolized**) in our bodies, the chemical energy in the food is changed into

● *thermal energy* (heat) to keep us warm,

● *mechanical energy*, to help us move around and keep our heart and breathing muscles working,

● *electrical energy* in our nerves, to help us to think and respond to stimuli like heat or pain.

Questions

1 (a) Give one example in each case of a food which contains a high proportion of (i) fat, (ii) carbohydrate.
(b) What elements do fats and carbohydrates contain?
(c) Name two carbohydrates.
(d) What substances are produced when carbohydrates are broken down (metabolized)? (Hint: the carbohydrates act like fuels.)

2 Look closely at figure 1.
(a) Draw a pie chart showing the percentages that different fuels made to the world's energy requirements in 1990.
(b) Which fuel showed
(i) the largest increase in percentage between 1970 and 1990,
(ii) the largest decrease in percentage between 1970 and 1990,
(iii) the same percentage in both 1970 and 1990?
(c) Why do you think the world energy requirements have increased steadily since 1970?

3 (a) Describe what happens when
(i) water is heated in a saucepan,
(ii) ice cubes are made in a freezer,
(iii) coal burns,
(iv) margarine is heated in a saucepan,
(v) a steel poker is heated in a fire.
(b) Make a summary of the types of change which can occur when substances are heated. For example, one type of change is that solids melt.

2 Fossil Fuels

Fossils in coal show the plants from which the coal was formed

The most commonly used fuels are coal, oil and natural gas. These fuels are called **fossil fuels** because they have formed from the remains of dead animals and plants.

How did fossil fuels form?

When plants and animals die and decay they rot away completely. During this process, compounds containing carbon, hydrogen and oxygen in the decaying material react with oxygen in the air. The products are carbon dioxide and water.

Sometimes, the plants and animals cannot react with oxygen when they die. If this happens, the carbon compounds in the decaying material turn into energy rich substances like coal and oil. Three hundred million years ago the Earth was covered in forests and the sea was full of tiny organisms. When some of these living things died, they were covered with mud and protected from oxidation. Over millions of years these deposits were changed by bacteria and compressed by the layers of earth, and the sea above them. Coal has formed mainly from plants that grew on land. Oil and gas have formed mainly from plants and animals that lived in the sea.

Obtaining fossil fuels

The pictures below show how coal is obtained by **mining** whereas oil and gas are obtained by **drilling**.

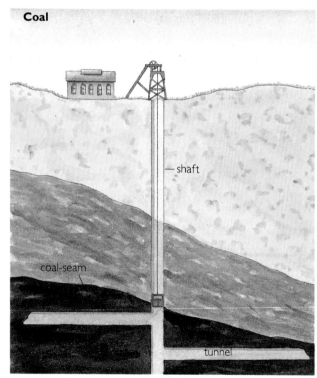

A mineshaft is sunk through layers of earth and rock to the coal seam so that the coal can be dug out

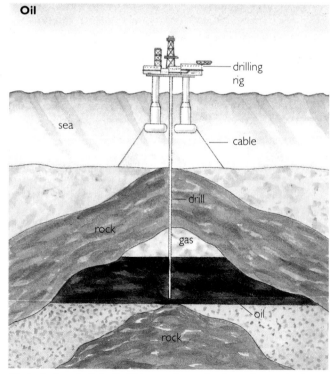

First of all, a rig is built. This may be on land, on the sea bed or anchored to the sea bed. Long drills are then used to bore through rock layers to the gas and oil. The pressure of the gas forces oil through pipes to the surface

Coal is mined all over the world. It is the largest source of fuel that we have, but the effects of coal mining often ruin the landscape. These photographs show the huge spoil heap near Dinnington Mine (South Yorkshire) and the same area after it was reclaimed

Conserving fossil fuels

Fossil fuels are concentrated stores of chemical energy. When they burn, some of their chemical energy is released as heat. For 150 years, industrial countries have relied on fossil fuels for energy. Vast amounts of coal, oil and natural gas have been used for heating, for transport and to generate electricity. Large amounts of fossil fuels are still being used for these purposes, but they cannot last forever. The Earth's resources of coal, oil and gas are limited. Eventually, they will run out. Figure 1 shows how long these fuels will last if we continue to use them as we do now.

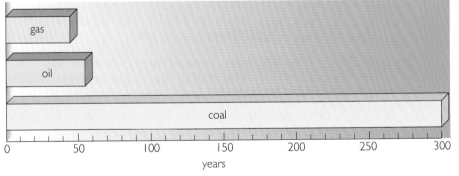

years

Figure 1
How long will the world's reserves of fossil fuels last?

Although coal is plentiful, oil and natural gas will probably start to run out in your lifetime. Because of this, it is important to conserve fossil fuels and avoid wasting energy. Our attempts to conserve fuels have led to:

- the more economic and more efficient use of fuels,

- the manufacture of special fuels for some uses,

- research into alternative energy sources to fossil fuels,

- the use of better materials and better methods for insulation.

Questions

1 (a) Draw a flow diagram to show how coal is formed.
(b) Why is coal described as a fossil fuel?
(c) Why is good ventilation important when coal burns?

2 Make a table showing how people have tried to conserve (save) fossil fuels. The table has been started for you.

Action taken	How it saves fossil fuels
Driving smaller cars . . .	Uses less petrol . . .

3 What problems does coal mining pose for (i) the safety of miners, (ii) the environment?

3 Alternative Energy Sources

Only a few tidal power stations have been built. This photo shows a tidal barrage in Nova Scotia, Canada

The CEGB's first wind turbine which is located at the Carmarthen Bay Power Station in South Wales. The 24 m high machine can supply up to 200 kW of electricity—enough to operate 200 one-bar electric fires

At one time, almost all the energy we used in our homes and industries came from fossil fuels. We still use vast amounts of coal, oil and natural gas, but these fossil fuels will not last forever. Because of this, scientists and engineers are looking for and developing **alternative** (other) **energy sources**.

- **Tidal power.** A tidal power station traps the high tides behind a barrage across a river estuary. When the tide falls, the water is made to flow through turbines which generate electricity.

- **Nuclear power.** Nuclear power is used to generate electricity. Uranium is used as the fuel. Nuclear energy has many advantages, but some people think that it is unsafe. Nuclear energy is discussed further in section K.

- **Hydro-electric power.** Falling water can be used to drive turbines which generate electricity. Once the power station has been built, the cost of hydro-electricity is fairly cheap. But because the process relies on water falling from a height, the use of hydro-electric generators is limited.

- **Solar power.** The Sun provides us with energy in two forms—heat and light. The Sun's thermal energy (heat) can be used to heat water in solar panels, or it can be used in solar furnaces. Light from the Sun can be used to make electricity using photo-cells.

- **Wind power.** Windmills, like water mills, have been used as a source of power for hundreds of years. Some scientists believe that giant windmills (or wind turbines as they are called) could be used to generate electricity on a large scale in the future.

Energy resources—savings and income

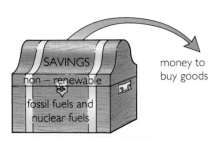

Figure 1
Non-renewable energy sources are like life savings—once they are spent, they can never be replaced

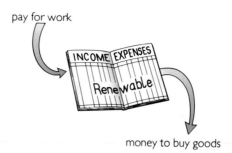

Figure 2
Renewable energy sources are like a regular income from a job—you can keep spending and the money is continually replaced by your wages

The Earth's energy resources can be compared to money. Fossil fuels and nuclear fuels are like *life savings* (figure 1). They represent energy stored (saved) over millions of years. They are sometimes called **non-renewable energy sources**. Once used, these fuels (like savings) are gone forever. We cannot get them back and use them again.

Fortunately, there are some **renewable energy sources**. Every day the Sun shines (somewhere!), the wind blows, rain falls and tides come in. So solar power, wind power, hydro-electric power and tidal power are renewable energy sources. We can use them all the time, but they keep on being replaced. They are like an *income from a job* (figure 2). In the future, we may have to use these renewable energy sources more and more as our fossil fuel savings run out.

Solar panels outside Foula lighthouse, Shetland

Questions

1 Draw a large sketch map of the British Isles. (You might like to copy or trace from an atlas.) Mark on your map (with a letter T) those places where tidal power might be used. Use the letters H, S and W to indicate areas where hydro-electric, solar and wind power might be used.

2 Get into a group with 2 or 3 others. Try to imagine how people in the UK might obtain the energy they need between now and the year 3000. Make a timetable with dates and important energy sources.

3 There are economic and environmental reasons for choosing different energy sources. What are the advantages and disadvantages of the following energy sources?
coal, tidal power, wind power.

4 (a) What is meant by (i) renewable energy sources, (ii) non-renewable energy sources?
(b) Is wood a renewable or a non-renewable energy source? Explain your answer.
(c) Is wave power a renewable or non-renewable energy source? Explain your answer.

5 Why is electricity much cheaper in Norway than in Britain?

4 Energy Transfers

Most of the energy that we needed for heating, for transport and for generating electricity comes from the chemical energy stored in fuels, foods and other materials.

Figure 1 summarizes the most useful ways in which chemical energy is converted into other forms of energy.

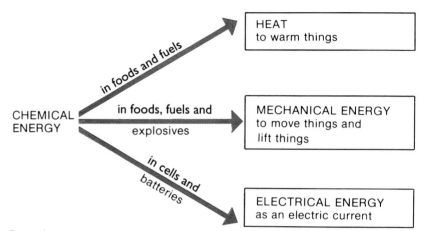

Figure 1
When you turn on a gas fire, digest a meal or use a torch, chemical energy changes into other forms of energy. But when you do these things, no energy is ever lost. It simply changes from one form to another

> This is summarized in the **Law of Conservation of Energy** which says: Energy cannot be destroyed. It can only be changed from one form to another.

Energy is most useful when it is concentrated. This is why fuels and foods are so useful. They store energy in concentrated form until it is needed. When the fuel is burnt or the food is eaten, the energy is released.

Respiration—Energy from Foods

Foods are broken down in our bodies by a chemical process called **respiration** (unit B3). During respiration, foods react with oxygen forming carbon dioxide and water. This is why we breathe out carbon dioxide and water vapour and why urine is mainly water. *Respiration is also exothermic* and energy is given out. Because of this, foods are sometimes described as 'biological fuels'.

food   +   oxygen → carbon + water + energy
(containing C and H)   dioxide

Photosynthesis—Energy from the Sun

Humans and other animals cannot make their own food. To get their food, they must eat other animals or plants. But plants are different. They can make their own food from carbon dioxide in the air and water in the soil. This process involves **photosynthesis** in which carbon dioxide and water are converted into carbohydrates like glucose ($C_6H_{12}O_6$), sugar ($C_{12}H_{22}O_{11}$) and starch. Oxygen is also produced (figure 2).

$$6CO_2 + 6H_2O \xrightarrow{sunlight} C_6H_{12}O_6 + 6O_2$$

The energy that Sally Gunnell needs to hurdle is provided by food

Figure 2
Photosynthesis

Photosynthesis is an endothermic process. The energy needed for the reaction comes from sunlight. Light energy from the sun is absorbed by chlorophyll, a green substance in the leaves of plants. Without the sun, photosynthesis would not occur and life on Earth would be impossible.

Ultimately, the sun is our major source of energy. Carbohydrates like glucose and starch, which are produced during photosynthesis, can be used by plants as foods. In turn, the plants can be used as foods for animals. Some plants such as potatoes, wheat, rice and sugar beet are grown as a source of food for humans and animals.

Photosynthesis and respiration are very complex processes involving several steps. Overall, photosynthesis is the reverse of respiration. Respiration is exothermic. It uses up food and oxygen and produces carbon dioxide, water and energy. In contrast, photosynthesis is endothermic. It uses up carbon dioxide, water and energy in sunlight to produce food and oxygen.

The two processes can be summarized as:

$$C_6H_{12}O_6 + 6O_2 \underset{\text{photosynthesis}}{\overset{\text{respiration}}{\rightleftharpoons}} 6CO_2 + 6H_2O + \text{energy}$$

Respiration happens in both plants and animals, but photosynthesis can only happen in plants. Respiration and photosynthesis are important in the **carbon cycle** (figure 3). This shows how carbon, in carbon dioxide and in other carbon compounds, is recycled through various processes.

Chlorophyll is contained in chloroplasts in plants. The chloroplasts can be identified easily in these spirogyra as bright zig-zags across the cells

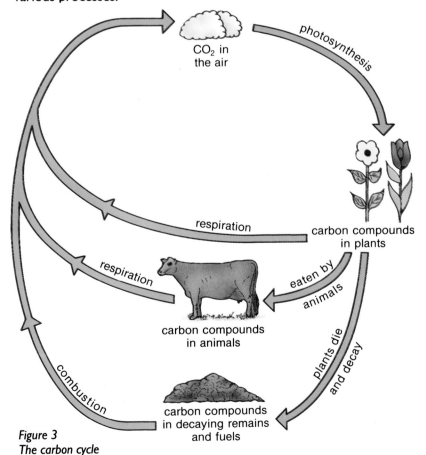

Figure 3
The carbon cycle

Questions

1 Explain the following:
respiration; *photosynthesis*; *carbon cycle*; *exothermic*.

2 (a) How do our bodies use the energy given out during respiration?
(b) Why are animals unable to photosynthesize?
(c) In what ways is photosynthesis the opposite of respiration?

3 *True or false?*
The reaction represented as:
$C_6H_{12}O_6 + 6O_2 \rightarrow 6CO_2 + 6H_2O$
A is exothermic.
B is reversible.
C involves a hydrocarbon.
D is called photosynthesis.
E involves several steps.
F involves chlorophyll.

4 Discuss the following statement with 2 or 3 others. 'Without the sun, life would be impossible. So, we should worship the sun'. Write a short report of your discussion.

5 Crude Oil

An aerial view of an oil refinery near Stavanger, Norway. Notice the fractionating column, the storage tanks and the jetty at which tankers can berth

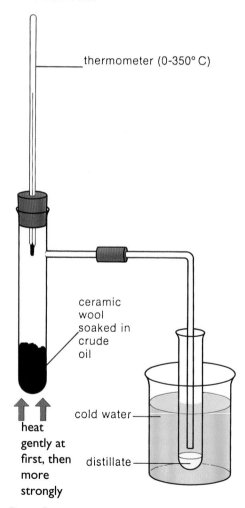

Figure 1
The small scale fractional distillation of crude oil

Crude oil (petroleum) is the main source of fuel and organic chemicals in the UK. The crude oil comes to our refineries, from the North Sea and the Gulf area in the Middle East. It is a sticky, smelly, dark-brown liquid. Crude oil is a mixture containing hundreds of different compounds, from simple substances like methane (CH_4) to complicated substances with long chains and rings of carbon atoms. Nearly all the substances in crude oil contain carbon. These carbon compounds are often called **organic compounds**.

A century ago oil was almost unknown. Now we could hardly survive without it. It is almost as important to our lives as air and water. In the UK, 70% of all organic chemicals come from oil. Antifreeze, brake fluid, lipstick, nylon, explosives and paint are all made from it. You may be dressed entirely in oil-based textiles, like Terylene or nylon. Without oil, most transport would come to a standstill and any machine larger than a toy car would seize up from lack of lubricant. We have enough oil to last another 60 years and new reserves are being discovered all the time. Even so, we are using up oil reserves so fast that we need to use crude oil more economically.

Boiling range	20–70°C	70–120°C	120–170°C	170–240°C
Name of fraction	Petrol	Naphtha	Paraffin	Diesel oil
Colour	Pale yellow	Yellow	Dark yellow	Brown
Viscosity	Runny	Fairly runny	Fairly viscous	Viscous
How does it burn?	Easily, with clean yellow flame	Quite easily, yellow flame, some smoke	Harder to burn, quite smoky flame	Hard to burn, smoky flame

Table 1: The properties of fractions obtained by the small-scale fractional distillation of crude oil

Separating crude oil by fractional distillation

At oil refineries, crude oil is separated into fractions by fractional distillation (unit A8). Then it is treated and purified to produce different fuels and chemicals. Figure I shows the small scale fractional distillation of crude oil. The ceramic wool, soaked in crude oil, is heated very gently at first and then more strongly so that the distillate slowly drips into the collecting tube. Four fractions are collected, with the boiling ranges and properties shown in table I. Table I also shows the industrial fractions to which our fractions correspond. Notice how the properties of the fractions gradually change in colour, viscosity and burning.

Constituents and uses of the different fractions

The fractions from crude oil contain mixtures of similar substances. These fractions contain compounds with roughly the same number of carbon atoms. Table 2 shows what each fraction contains and what it is used for. The uses of the fractions depend on their properties. Petrol vaporizes easily and is very flammable, so it is ideal to use in car engines. Lubricating oil, which is very viscous, is used in lubricants and in central heating. Bitumen, which is solid but easy to melt, is used for waterproofing and asphalting on roads.

Bitumen is mixed with stone chippings and used to surface roads

Fraction	Boiling range	Number of carbon atoms in the constituents	Uses
Fuel gas	−160 to 20°C	1–4. Methane (CH_4), ethane (C_2H_6), propane (C_3H_8) and butane (C_4H_{10})	Fuel for gas ovens, LPG, GAZ, chemicals
Petrol (gasoline)	20 to 70°C	5–10. e.g. octane (C_8H_{18})	Fuel for gas vehicles, chemicals
Naphtha	70 to 120°C	8–12.	Chemicals
Paraffin (Kerosine)	120 to 240°C	10–16.	Fuel for central heating, jet engines and chemicals
Diesel oils and lubricating oils	240 to 350°C	15–70.	Fuel for diesel engines, trains and central heating, chemicals, lubricants
Bitumen	above 350°C	More than 70.	Roofing, waterproofing, asphalt on roads

Table 2: The constituents and uses of fractions from crude oil

Questions

1 Look at the results in table 1. How do the properties of the fractions change?
2 (a) Why is crude oil important?
(b) Why should we try to conserve our reserves of crude oil?
(c) How is crude oil separated into various fractions at a refinery?
(d) What should we do to ensure that crude oil is used more economically?
3 (a) List the main fractions obtained from crude oil.
(b) Give the main uses of each fraction.
4 (a) What does the word 'organic' mean in everyday use?
(b) Why do you think carbon chemistry is called 'organic chemistry'?
(c) Organic compounds are simple molecular compounds. What physical properties would you expect them to have?

6 Alkanes

There are millions of different carbon compounds. Carbon can form millions of different compounds because carbon atoms can form strong covalent bonds with each other. Atoms of other elements cannot do this. Remember diamond and graphite—giant structures of carbon atoms joined together by strong covalent bonds. Because of these strong C—C bonds, carbon forms molecules containing long chains of carbon atoms. There are thousands of compounds containing just hydrogen and carbon. These are called **hydrocarbons**.

The four simplest hydrocarbons are methane, ethane, propane and butane. Figure 1 shows the molecular formulas, structural formulas and molecular models for these four hydrocarbons. The structural

Name	methane	ethane	propane-	butane
Molecular formula	CH_4	C_2H_6	C_3H_8	C_4H_{10}
Structural formula	H | H — C — H | H	H H | | H — C — C — H | | H H	H H H | | | H — C — C — C — H | | | H H H	H H H H | | | | H — C — C — C — C — H | | | | H H H H

Figure 1

A model of methane

A model of ethane

A model of propane

A model of butane

formulas show which atoms are attached to each other. But they cannot show the correct three-dimensional structure of the molecules. The 3-D structures are shown in the molecular models. There are four covalent bonds to each carbon atom. Each of these bonds consists of a pair of electrons shared by two atoms. The four pairs of electrons around a carbon atom repel each other as far as possible. So, the bonds around each carbon atom spread out tetrahedrally, as in diamond (figure 2).

Methane, ethane, propane and butane are members of a series of compounds called **alkanes**. All other alkanes are named from the number of carbon atoms in one molecule. So C_5H_{12} is *pentane*, C_6H_{14} is *hexane*, C_7H_{16} is *heptane* and so on. The names of all alkanes end in *-ane*.

Look at the formulas of methane, CH_4, ethane, C_2H_6, propane, C_3H_8 and butane C_4H_{10}. Notice that the difference in carbon and hydrogen atoms between methane and ethane is CH_2. The difference between ethane and propane is CH_2 and the difference between propane and butane is also CH_2. This is an example of a **homologous series**—a series of compounds with similar properties in which the formulas differ by CH_2.

Isomerism

Earlier in this unit, we noted that carbon could form a large number of compounds. Even more compounds are possible than we might expect, because it is sometimes possible to join the same set of atoms in different ways. For example, take the formula C_4H_{10}. We know already that butane has the formula C_4H_{10}. But there is another compound which also has this formula. This other compound is called methylpropane (figure 3). Notice that each carbon atom in butane and in methylpropane has 4 bonds and each H atom has 1 bond. Butane and methylpropane are two distinct compounds with different properties (see figure 3). Compounds like this with the same molecular formula, but different structural formulas but different properties are called **isomers**.

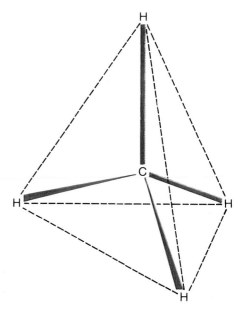

Figure 2
The tetrahedral arrangement of bonds in methane

Name	butane	methylpropane
Molecular formula	C_4H_{10}	C_4H_{10}
Structural formula	$H-\overset{\overset{H}{|}}{\underset{\underset{H}{|}}{C}}-\overset{\overset{H}{|}}{\underset{\underset{H}{|}}{C}}-\overset{\overset{H}{|}}{\underset{\underset{H}{|}}{C}}-\overset{\overset{H}{|}}{\underset{\underset{H}{|}}{C}}-H$	$H-\overset{\overset{H}{|}}{\underset{\underset{H}{|}}{C}}-\overset{\overset{H-\overset{H}{|}-H}{|}}{\underset{\underset{H}{|}}{C}}-\overset{\overset{H}{|}}{\underset{\underset{H}{|}}{C}}-H$
Melting point/°C	−138	−159
Boiling point/°C	0	−12
Density /g cm⁻³	0.58	0.56

Figure 3
Isomers with the formula C_4H_{10}

Questions

1 Explain the following:
hydrocarbon; alkane; homologous series; isomerism.

2 (a) What is C_8H_{18} called?
(b) Draw a structural formula for C_8H_{18}.
(c) How many H atoms will the alkane with 10 carbon atoms have?
(d) If an alkane has *n* carbon atoms, how many H atoms will it have?

3 Why can a homologous series of compounds be compared to a group of elements in the periodic table?

4 Draw the structural formulas for (i) the two isomers with the formula C_3H_7Cl; (ii) the two isomers with the formula C_2H_6O; (iii) the three isomers with the formula C_5H_{12}.

Alkanes are typical molecular (non-metal) compounds. Some of their properties are described below.

Volatility

Tar and bitumen contain alkanes with a relative molecular mass of more than 500. Even so, they begin to melt on very hot days. Compare this with an ionic compound like sodium chloride. NaCl has a relative formula mass of only 58.5, yet its melting point is 808° C.

Crude oil and natural gas are the main sources of alkanes. Alkanes with up to four carbon atoms in each molecule are gases at room temperature. Methane (CH_4) and ethane (C_2H_6) are the main constituents of natural gas. Propane (C_3H_8) and butane (C_4H_{10}) are the main constituents of natural gas. Propane (C_3H_8) and butane (C_4H_{10}) are the main constituents of 'liquefied petroleum gas' (LPG). The best known uses of LPG are 'Calor gas' and GAZ for camping, caravans and boats. Butane is also used as lighter fuel and in portable hair tongs.

A small blue butane cylinder used to fuel a gas barbecue

Two red propane cylinders, stored outside, and used to fuel a domestic gas cooker

Alkanes with 5 to 17 carbon atoms are liquids at room temperature. Mixtures of these liquids are used in petrol, in paraffin and in lubricating and engine oils.

Alkanes with 18 or more carbon atoms per molecule are solids at room temperature. Notice that the alkanes become less volatile and change from gases to liquids and then to solids as their molecular size increases.

Insoluble in water

Alkanes are insoluble in water, but they dissolve in organic solvents such as tetrachloromethane (carbon tetrachloride) and petrol.

Poor reactivity

Alkanes do not contain ions and their C—C and C—H bonds are very strong. So they have very few reactions. They do not react with metals, aqueous oxidizing agents, acids or alkalis. It may surprise you, but petrol (which contains mainly alkanes) will not react with sodium, potassium permanganate or concentrated sulphuric acid. You must **not** try any experiments with petrol. It is highly flammable.

Combustion

The most important reactions of alkanes involve combustion. They burn in oxygen, producing carbon dioxide and water.

$$C_4H_{10} + 6\tfrac{1}{2}O_2 \rightarrow 4CO_2 + 5H_2O$$
butane in GAZ

The combustion reactions are very exothermic, so alkanes in natural gas and crude oil are used as fuels. When there is too little oxygen for combustion, carbon (soot) and carbon monoxide form as well as carbon dioxide. Carbon monoxide is very toxic, so it is dangerous to burn carbon compounds in a poor supply of air.

Concorde being refuelled. Kerosine (paraffin) is used as the fuel in aircraft

Reaction with halogens

Chlorine and bromine are even more reactive non-metals than oxygen and halogens will also react with alkanes. Ethane reacts with bromine to form bromoethane and hydrogen bromide.

This reaction is further illustrated with molecular models below.

One of the H atoms in ethane is substituted by a bromine atom forming bromoethane. This is called a **substitution reaction**. If there is enough bromine, more hydrogen atoms can be replaced by bromine atoms. Chlorine reacts in a similar way to bromine. The reactions of alkanes with chlorine and bromine are very slow at room temperature, but they go much faster when the mixture is heated or exposed to ultraviolet light.

Questions

1 (a) What is meant by a substitution reaction?
   (b) Write an equation for the reaction of methane with chlorine.
   (c) Write the formulas of all possible substitution products when methane reacts with chlorine.
2 (a) Write an equation for the complete combustion of octane in petrol.
   (b) What are the products when butane burns in a poor supply of oxygen?
   (c) Why is it dangerous to allow a car engine to run in a garage with the door closed?
3 Why do alkanes change from gases to liquids and then to solids as their molecular size increases?
4 (a) Use molecular models to build a molecule of C_6H_{14}.
   (b) What is C_6H_{14} called?
   (c) Draw a structural formula for C_6H_{14}.

8 Petrol from Crude Oil

Fraction	Approximate % in	
	crude oil	everyday demand
Fuel gas	2	4
Petrol	6	22
Naphtha	10	5
Kerosine	13	8
Diesel oil	19	23
Fuel oil and bitumen	50	38

Relative amounts of different fractions in crude oil and in the demand for each fraction

After 1930 the demand for petrol increased much faster than the demand for heavier fractions which make up three quarters of crude oil. This meant that refineries were left with large surpluses of the heavier fractions. Fortunately, chemists have found ways of converting the heavier fractions into petrol and other useful products.

One method of breaking up the large alkane molecules in these heavier fractions is **catalytic cracking**.

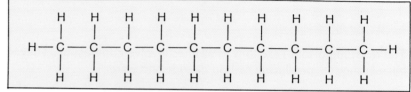

Figure 1 The structural formula of decane

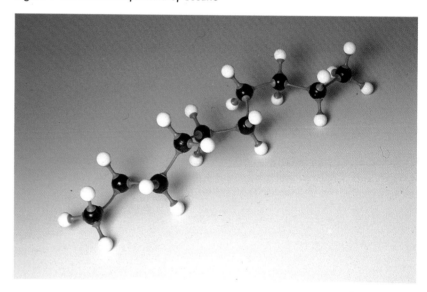

A molecular model of decane

Why is unleaded cheaper than four-star?

What are the products of cracking?

Look at the long molecule of decane ($C_{10}H_{22}$) in figure 1. It has a chain of ten carbon atoms with twenty-two hydrogen atoms. Imagine that decane is cracked (split) between two carbon atoms. This cannot produce two smaller alkane molecules because there are not enough hydrogen atoms to go round. But suppose that one product is the alkane, octane (C_8H_{18}). If C_8H_{18} is split off from $C_{10}H_{22}$, the molecular formula of the remaining part is C_2H_4 (figure 2). The chemical name for C_2H_4 is **ethene**. Notice in figure 2 that ethene has a *double* bond between the two carbon atoms. This double bond allows all the carbon atoms in the products to have four bonds.

Hydrocarbons such as ethene, which contain a double bond (C=C), are known as **alkenes**. Their names come from the alkane with the same number of carbon atoms, using the ending *-ene* rather than *-ane*. Organic compounds, like alkanes, which have four single covalent bonds to all their carbon atoms, are described as **saturated compounds**. Alkenes, which have double bonds between some carbon atoms, are examples of **unsaturated compounds**.

$$C_{10}H_{22} \qquad\Rightarrow\qquad C_8H_{18} \quad + \quad C_2H_4$$

decane $\qquad\Rightarrow\qquad$ octane $\quad+\quad$ ethene

Figure 2
When decane undergoes catalytic cracking, it forms an alkane, like octane, and an alkene, like ethene.

How are molecules cracked?

Unlike distillation, cracking is a chemical process. It involves breaking a strong covalent bond between two carbon atoms. This requires high temperatures and a catalyst. At high temperatures the larger alkane molecules have more energy and they break apart into two or more smaller molecules. The catalyst is finely-powdered aluminium oxide and silicon(IV) oxide. These substances do not react with the crude oil fractions but they do provide a hot surface that speeds up the cracking process.

Cracking is important because it helps to produce more petrol. Larger alkanes in crude oil are cracked to produce alkanes with about 8 carbon atoms like octane, the main constituent in petrol. The petrol obtained in this way is better quality than that obtained by the distillation of crude oil. Cracked petrol is therefore blended with other petrols to improve their quality.

The catalytic cracking plant at the Wilmington refinery, Connecticut, USA

Questions

1 Explain the words:
cracking; alkene; unsaturated.
2 (a) Why is cracking important?
  (b) What conditions are used for cracking?
  (c) Write an equation for the cracking of octane in which one of the products is pentane.
  (d) Draw the structural formulas of the products in the equation in (c).
3 (a) What is the name of the alkene of formula C_3H_6?
  (b) Draw the structure of this alkene.
  (c) Why is there no alkene called methene?
4 (a) Design a simple experiment in which you could attempt to crack some kerosine (paraffin) in order to obtain ethene. (Ethene is a gas at room temperature.)
  (b) What safety precautions will you take in your experiment?
  Do not carry out your experiment until it has been checked by your teacher.
5 What changes do you think there will be in our lives when crude oil begins to run out?

9 Ethene

Ethene is an important industrial chemical. It is manufactured by cracking the heavier fractions from crude oil. It can be made on a small scale by cracking paraffin oil using the apparatus in figure 1. **Wear eye protection** if you are preparing ethene in this way. Remember that both paraffin and ethene are flammable. Beware of suck back and take care that the delivery tube does not get blocked. Heat the middle of the tube below the porous pot or aluminium oxide. Heat will be conducted along the tube to vaporize the paraffin. The main gaseous product is ethene. Figure 2 shows some of the properties of ethene.

In nature, ethene acts as a trigger for the ripening of fruit, particularly bananas

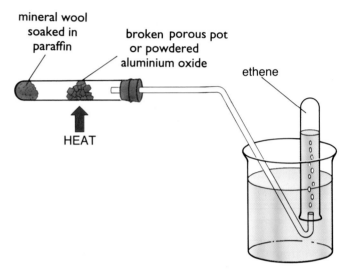

mineral wool soaked in paraffin
broken porous pot or powdered aluminium oxide
ethene
HEAT

Figure 1 Preparing ethene by cracking paraffin oil

Reactions of alkenes

Alkenes, such as ethene, are much more reactive than alkanes. The most stable arrangement for the four bonds to a carbon atom is a tetrahedral one with four *single* bonds. This means that a C=C bond is unstable. Other atoms can add across the double bond to make two single bonds. So, alkenes readily undergo **addition reactions**.

This explains why ethene decolorizes bromine water and acidified potassium permanganate solution. When ethene is shaken with bromine water the yellow colour disappears. The bromine molecules add across the double bond in ethene forming 1,2-dibromoethane.

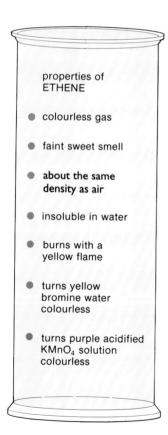

properties of
ETHENE

● colourless gas

● faint sweet smell

● **about the same density as air**

● insoluble in water

● burns with a yellow flame

● turns yellow bromine water colourless

● turns purple acidified $KMnO_4$ solution colourless

Figure 2

$$ \underset{\text{ethene}}{\begin{array}{c} H \\ \diagdown \\ C = C \\ \diagup \diagdown \\ H H \end{array}} + \underset{\text{bromine}}{Br_2} \rightarrow \underset{\text{1,2-dibromoethane}}{\begin{array}{c} H H \\ | | \\ H-C-C-H \\ | | \\ Br Br \end{array}} $$

1,2-dibromoethane is an important additive in petrol. Tetraethyllead(IV), $Pb(C_2H_5)_4$, is also added to petrol to help it burn smoothly and prevent 'knocking'. Unfortunately the tetraethyllead(IV) causes lead to be deposited in the cylinders and round the sparking plugs. 1,2-dibromoethane reacts with the deposit of lead to form volatile lead bromide that passes out with the exhaust gases.

Ethene also has an addition reaction with hydrogen at 150° C using a nickel catalyst. The product of this reaction is ethane.

This process is known as **catalytic hydrogenation**. It is important in making lard and margarine from vegetable oils in palm seeds and sunflower seeds. The vegetable oils are liquids containing alkenes. During hydrogenation these alkenes are converted to alkanes. This change in structure can turn an oily liquid into a harder, fatty solid that can be used to make margarine.

Compounds, like ethene, which contain double bonds are described as unsaturated. *Compounds, like ethane, with only single bonds are* saturated *hydrocarbons. Which of the products in this photo is the most unsaturated?*

By controlling the amount of hydrogen added to vegetable oils, margarine can be made as hard or as soft as required. Soft margarine is being made in this factory

Comparing ethene with ethane

Ethane and other alkanes are fairly unreactive. They react only with oxygen (burning) and other reactive non-metals like chlorine and bromine. They will not react with bromine water or dilute potassium permanganate $KMnO_4$) solution.

In comparison, ethene and alkenes are very reactive because of the addition reactions that take place across their C=C bonds. They readily decolorize bromine water and dilute potassium permanganate and these reactions are used to test for alkenes.

Questions

1 (a) Explain what 'substitution reaction' and 'addition reaction' mean.
  (b) Use these words to compare the way in which ethane and ethene react with bromine.
2 (a) Write the structural formula for propene, C_3H_6.
  (b) Write equations for the reactions of propene with (i) hydrogen; (ii) bromine.
  (c) Draw the structural formulas of the products in part (b).
3 Three cylinders of gas are known to contain ethane, ethene and carbon dioxide but their labels are not clear. What simple tests would you make to show which gas is which?
4 Why are alkenes much more reactive and alkanes?
5 Why do some people object to the addition of tetraethyllead(IV) to petrol?
6 (a) How is margarine made from vegetable oils?
  (b) Explain what happens during the reaction in (a).
  (c) How can the melting point of the margarine be controlled?

10 Products from Ethene

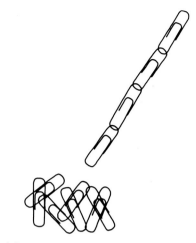

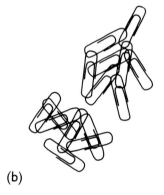

(a)

(b)

Figure 1
(a) Correct conditions for polymerization;
(b) Unsuitable conditions for polymerization

POLYTHENE
- tough
- light
- flexible
- easily moulded
- transparent
- easily coloured
- good insulator
- no reaction with water, acids or alkalis

Table 1: Properties of polythene

Ethene is one of the most valuable feedstocks for the chemical industry. Because it is so reactive, ethene is used to make a number of useful products including

1 Polythene (polyethene)

2 PVC (polychloroethene)

3 Ethanol (methylated spirits)

4 1,2-dibromoethane (see unit 9).

Clingfilm is made from polythene

- **Polythene (polyethene).** In the last unit we saw that molecules of ethene can have addition reactions with other substances such as hydrogen and bromine. If the conditions are right, molecules of ethene will also add to each other. Hundreds of ethene molecules may join together, forming a giant molecule of polythene.

$$n \left(\begin{array}{c} H \\ \diagdown \\ C = C \\ \diagup \\ H \end{array} \begin{array}{c} H \\ \diagup \\ \diagdown \\ H \end{array} \right) \xrightarrow[+ \ catalyst]{high \ temp, \ high \ pressure} \left(\begin{array}{cc} H & H \\ | & | \\ -C - C - \\ | & | \\ H & H \end{array} \right)_n$$

In the structure of polythene, n is between 500 and 1500. This process, in which small molecules add to each other to form a giant molecule, is called **polymerization**. The giant molecule is called a **polymer**. The small molecules, like ethene which add to each other are called **monomers**. It is essential that the conditions used for polymerization are correct. Figure 1 shows a model (using paper clips) of what happens when unsuitable conditions are used.

Table 1 shows the properties of polythene. These have led to many different uses and polythene is the most important plastic at present. It is used as thin sheets for packing and coating materials. It is moulded into beakers, buckets and troughs. It is made into pipes and used to insulate underwater cables.

- **PVC (polychloroethene).** PVC is almost as widely used as polythene. PVC fabric is made into rainwear, wallpaper, curtains, furniture upholstery and protective clothing in industry. Rigid PVC is used for records and for gas and water pipes. Flexible PVC is used for toys and for insulating cables. PVC is made by the polymerization of chloroethene which itself comes from ethene.

$$n\left(\begin{array}{c} H \\ \backslash \\ C = C \\ / \\ H \end{array}\begin{array}{c} H \\ / \\ \\ \backslash \\ Cl \end{array}\right) \xrightarrow{polymerization} \left(\begin{array}{cc} H & H \\ | & | \\ -C-C- \\ | & | \\ H & Cl \end{array}\right)_n$$

All plastics (like polythene, PVC, perspex and polystyrene) and manmade fibres (like nylon and Terylene) are polymeric materials. The main raw material for all these products is crude oil. Although these polymers have provided us with many new materials, they have one big disadvantage. They are **non-biodegradable**. This means that they are not decomposed by the weather or by bacteria, like paper, wood and steel. As litter, they often pollute the environment.

The upper walls and roof of this house have an outer covering of PVC

- **Ethanol.** Ethanol is manufactured by the addition of water to ethene. The reaction is carried out at 300° C and very high pressure using a catalyst of phosphoric acid.

$$\begin{array}{c} H \\ \backslash \\ C = C \\ / \\ H \end{array}\begin{array}{c} H \\ / \\ \\ \backslash \\ H \end{array} + H_2O \rightarrow \begin{array}{c} H\ H \\ | \ | \\ H-C-C-H \\ | \ | \\ H\ OH \end{array}$$

ethane   water   ethanol

Ethanol is the main constituent of methylated spirits, the most widely-used industrial solvent. Methylated spirits is used as a solvent for paints, resins, soaps and dyes. Pure ethanol is used as a solvent for perfumes, cosmetics and after-shave lotions.

Questions

1 Explain the following:
polymer; addition polymerization; non-biodegradable.
2 PVC is made by the polymerization of chloroethene ($CHCl=CH_2$). Draw simple diagrams to show the structure of: (i) a molecule of chloroethene; (ii) a section of the PVC molecule.
3 (a) How is ethene manufactured?
(b) Why is ethene important in industry?
4 *True* or *false?*
Polyethene
A has the same empirical formula as ethene.
B has the same molecular formula as ethene.
C is an alkane.
D is an alkene.
E is a hydrocarbon.
F undergoes addition reactions like ethene.
5 Give one use of and one disadvantage of plastic rubbish.

11 Making Plastics

Polythene sheeting emerging from an extrusion machine after manufacture

Plastics are made by polymerization. There are two types of polymerization:

- **additional polymerization,**
- **condensation polymerization.**

Addition polymerization

Polythene, PVC, polystyrene and perspex (acrylic) are all made by addition polymerization.

Polythene is made by heating ethene at high pressure with special catalysts. Under these conditions the ethene molecules add together (polymerize). Ethene molecules have a double bond between two carbon atoms. During polymerization, these double bonds 'open up' and the carbon atoms on separate ethene molecules join together to form a molecule of polythene. Polythene is short for poly(ethene). The name polythene is used because 'poly' means many. So polythene means 'many ethenes' have been joined together.

ethene molecules (monomers) → part of the polythene (molecule polymer)

When ethene forms polythene, very long chains are produced. These long molecules contain anywhere between 500 and 50 000 carbon atoms.

Three other plastics formed by addition polymerization are listed in table 1. The table also shows the structure of the monomers which are used to make each polymer.

Recycled polythene is used in industries where the toxicity of plastic is not a critical factor

Plastic (polymer)	Monomer	
PVC	vinyl chloride (chloroethene)	H, Cl, $C{=}C$, H, H
Polystyrene	Styrene	H, C_6H_5, $C{=}C$, H, H
Perspex (acrylic)	methylmethacrylate	H, $C{=}C$, $O{-}C$, O, CH_3, H, H

Table 1: Some plastics and the monomers used to produce them

Condensation polymerization

Nylon and Terylene are made by condensation polymerization. These polymers are made from two different monomers, not just one like addition polymers. The two different monomers have reactive atoms at the end of their molecules. These react with each other joining the two monomers together. For example, Nylon is made from

diaminohexane

```
     H       H  H  H  H  H  H       H
      \      |  |  |  |  |  |      /
       N—C—C—C—C—C—C—N
      /      |  |  |  |  |  |      \
     H       H  H  H  H  H  H       H
```

and

hexanedioyl-
dichloride

```
     O       H  H  H  H       O
      \\      |  |  |  |      //
        C—C—C—C—C—C
      /      |  |  |  |      \
     Cl      H  H  H  H       Cl
```

These complex molecules can be shown much simpler as

H ———⬭——— H   and   Cl ———▭——— Cl

When one molecule of diaminohexane reacts with one molecule of hexanedioyldichloride, they join together. Hydrogen chloride is also produced.

But this reaction can occur again and again at the ends of the product, forming a very long polymer.

Figure 1 shows a method which you could use to make Nylon.

Most plastics consist of molecules with very long, thin chains. Sometimes, however, the long polymer molecules can form bonds with each other at points along the chain. This produces cross-linked three-dimensional structures which are harder and more rigid. Bakelite, melamine, resins and epoxyglues are examples of these plastics.

Once a plastic has been produced, it must then be turned into a useful article. This is usually done by

- **moulding** the warm, soft plastic under pressure, or

- **extruding**, or 'pushing out', the warm, soft plastic into different shapes by forcing it through nozzles or between rollers.

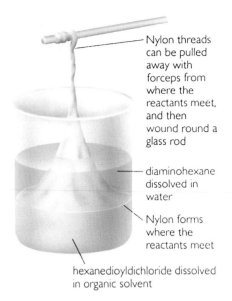

Nylon threads can be pulled away with forceps from where the reactants meet, and then wound round a glass rod

diaminohexane dissolved in water

Nylon forms where the reactants meet

hexanedioyldichloride dissolved in organic solvent

Figure 1
*Making Nylon. **Wear eye protection** if you try this experiment and remember the reagents are harmful*

Questions

1 (a) Use molecular models to make three ethene molecules.
(b) Using these three ethene molecules, break the double bonds and produce a short portion of polythene chain. (The equation on the opposite page will help you with this.)

2 Look at the structures of the monomers in table 1.
(a) Which part of their structures do they have in common?
(b) What do you think happens when these monomers polymerize to form polymers?
(c) Draw a short section of the polymer chain for PVC. Show six carbon atoms in the chain.

3 Find out what you can about the processes used to make everyday articles from plastics. Some processes to look out for are:
vacuum forming, extruding, moulding, foaming, thermoforming.

12 The Structure of Plastics

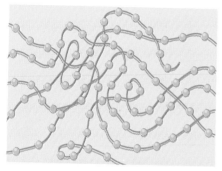

Figure 1
Tangled chains of long, thin molecules in a thermosoftening plastic

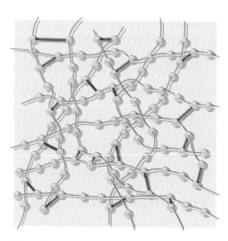

This plastic piping is made from a thermosoftening plastic. It has been warmed and then moulded into the required shape

Figure 2
Cross-links in a thermosetting plastic

Thermosoftening plastics

When polythene and Nylon are warmed, they become *soft* and can be moulded. As they cool, they set hard again. Because of this, they are called **thermosoftening plastics**. PVC, polystyrene and acrylic are also thermosoftening plastics.

Thermosoftening plastics contain long, thin molecules which form tangled chains (figure 1).

In these plastics there are strong forces between the atoms along each chain, but weak forces between one chain and another. This means that the chains can move over each other easily on stretching, flexing or warming.

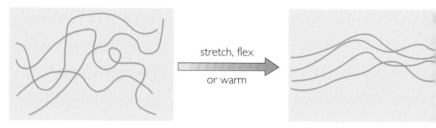

So, thermosoftening plastics

- stretch easily,
- are flexible,
- soften on warming,
- melt at low temperatures,
- can be shaped by warming and then moulding or extruding.

Thermosetting plastics

When Bakelite, melamine and resin are heated, they do *not* become soft. Once these plastics have been produced and moulded, they set hard and cannot be re-melted. Because of this, they are called **thermosetting plastics**. Urea-formaldehyde resin, polyester resin and epoxy glues are also thermosetting plastics.

Thermosetting plastics contain a network of large cross-linked molecules (figure 2).

In these cross-linked plastics, there are strong forces between the atoms along each chain *and* also strong cross-link forces between one chain and another. This means that the chains cannot move over each other on stretching or warming.

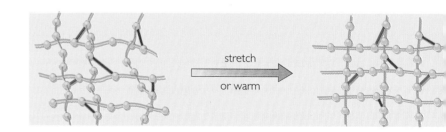

So, thermosetting plastics

- are hard and rigid,
- burn or char on heating,
- do not flex,
- do not soften or melt on heating.

Some thermosetting polymers form cross-links and harden at room temperature when a catalyst is added. One example of this is the polyester resin used in repair kits for car bodywork (see right). The polyester molecules form cross-links and harden when mixed with the catalyst (hardener).

Bakelite is a thermosetting plastic. It was one of the first plastics to be developed and used for everyday objects like this sugar jar. It is not used widely these days because it is relatively fragile and expensive to produce, compared with modern plastics

Isopon is used to repair car body work. It contains a polyester resin which forms cross links and sets hard when mixed with a catalyst

Questions

1 (a) Thermosoftening plastics can be compared to chocolate when it is warmed. Why is this?
(b) Thermosetting plastics can be compared to an egg when it is warmed. Why is this?
(c) What happens to the polythene molecules in clingfilm when it is stretched?

2 The properties of six important plastics are shown in the table below. Use the data in the table to suggest which plastic could be used to manufacture
(a) containers for lemon juice,
(b) laboratory safety screens,
(c) containers for soft cheese,
(d) the handles for screwdrivers,
(e) the inside surface of frying-pans,
(f) tea mugs.
In each case explain why you chose the particular plastic.

Plastic	Relative strength	Flexibility	Maximum temperature for use/°C	Resistance to dilute acids	Resistance to oils	Clarity	Cost /£ per kg
High density polythene	4	fairly stiff	150	excellent	good	poor	0.8
Low density polythene	1	very floppy	70	good	good	poor	0.7
Perspex	9	stiff	90	good	good	excellent	1.6
Polycarbonate	9	stiff	140	excellent	poor	good	3.0
PTFE	30	fairly flexible	250	excellent	excellent	poor	20.0
Urea-formaldehyde resin	9	very stiff	75	poor	good	poor	0.9

13 Plastic Waste

What should we do to prevent plastic rubbish from littering our streets and the countryside?

In some areas, plastic waste is a serious problem. Plastic bottles, polythene bags and other plastic articles litter our streets and the countryside. Farmers have even complained of cows choking and dying after swallowing plastic bags.

Unfortunately, the mess from plastic waste is made worse because most plastics are **non-biodegradable**. This means that they cannot be broken down by bacteria and the weather, like paper, wood and steel. So plastic rubbish lies around for years and years. In the last few years, biodegradable plastics have been developed. These are now used for bags and other wrappings.

Getting rid of plastic waste

There are 3 ways in which we deal with plastic waste at present.

● Landfill

Most plastic waste is dumped with other waste into rubbish tips.

A landfill site for refuse. Notice the large proportion of plastic waste

● Incineration

Plastics produce a lot of energy when they burn. So plastics are burnt in incinerators with other combustible waste. The heat produced can be used to heat homes and factories or to generate electricity.

Unfortunately, poisonous fumes are produced when some plastics are burnt. These fumes must be removed before the waste gases are released into the air.

● Recycling

About 2 million tonnes of plastic waste is produced in the UK each year. At present, only a small amount of this is recycled. Thermosoftening plastics are fairly easy to recycle. They can be melted or softened and then remoulded. This is shown on the left-hand side of figure 1.

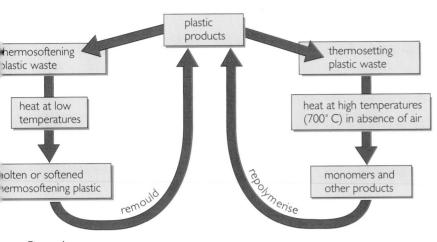

Figure 1
Recycling plastics

Thermosetting plastics are more difficult to recycle. They cannot be melted without decomposing. They are usually recycled by heating to about 700° C in the absence of air. This process is called **pyrolysis**. Under these conditions, the polymers cannot burn. Instead, they decompose to form their original monomers and other simple molecules. The original molecules can then be re-polymerized. This is shown on the right-hand side of figure 1.

At present, it is uneconomical to recycle most plastic waste. This is largely because of the cost of collecting and separating it, and the cost of energy in the recycling process. It is still cheaper to make new plastics from crude oil.

However, as oil becomes more scarce and more expensive, recycling will become more economical. This is why research into pyrolysis and other recycling processes is being carried out.

Recycling polythene

When foam-filled furniture burns, the fumes produced are very poisonous

Questions

1 A small packaging company produces 50 kg of plastic waste each week. This could be used in an incinerator to provide hot water and heating for the factory. The incinerator would cost £1000.

Material	Heat produced by burning 1 kg
Polythene	40 000 kJ
Polystyrene	40 000 kJ
PVC	20 000 kJ
Heating oil	40 000 kJ

(a) Use the table to calculate how many kilograms of heating oil the plastic waste could save each week. Explain your calculation.

(b) What would this save in the cost of heating oil per week? (Assume heating oil costs £2 per kg.)

(c) How long would it take the company to save enough money to pay off the cost of the incinerator?

2 If plastics are not burnt completely, poisonous substances are produced. Some plastics burn to produce acidic gases containing hydrogen chloride.

Design an incinerator in which plastics could be burnt completely. The incinerator must not expel any unburnt plastic or acidic gases.

3 Get into a small group with 2 or 3 others.

(a) Make a list of the main problems of recycling plastic waste.

(b) Make another list showing the advantages of recycling plastic waste.

14 Fermentation

Cider apples arriving in the silos at the Bulmer cider mill. Fifty thousand tonnes are processed each season

Making ethanol.

Nowadays, ethanol is manufactured from ethene (unit 110). At one time, however, ethanol and methylated spirits were manufactured by fermentation. Fermentation is, of course, still important in winemaking, brewing and bread making.

The starting material for fermentation is usually starch, sucrose or glucose. In wine making, a sweet sugary liquid is extracted from grapes. In beer making, starch is extracted from barley by soaking in hot water. The starch is then heated to 55° C with malt. The malt contains enzymes that break down the starch to maltose.

$$\text{starch} \xrightarrow[\text{in malt}]{\text{enzymes}} \text{maltose } (C_{12}H_{22}O_{11})$$

This reaction is very similar to the effect of saliva on starch.

Finally, yeast is added to the maltose. Yeast contains enzymes which break down the maltose first to glucose and then to ethanol and carbon dioxide.

$$\underset{\text{maltose}}{C_{12}H_{22}O_{11}} + H_2O \rightarrow \underset{\text{glucose}}{2C_6H_{12}O_6}$$

$$\underset{\text{glucose}}{C_6H_{12}O_6} \rightarrow \underset{\text{ethanol}}{2C_2H_5OH} + 2CO_2$$

The fermented liquid contains only 5 to 10% ethanol. Ethanol can be obtained from this liquid by fractional distillation.

In alcoholic drinks it is the ethanol from the fermentation that is important (making the drinks intoxicating). In bread making, the carbon dioxide from fermentation is important because this makes the bread rise before it is baked.

Alcoholic drinks containing ethanol, like beer and wine, can be pleasant in small amounts. They make people more relaxed and they have different tastes. But alcoholic drinks taken in excess can be very dangerous. They make us slow to react, they damage the liver and they can be addictive like drugs.

Foam in copper brewery tanks. What causes the foam?

Ethanol and alcohols

Ethanol is a member of a large class of compounds called alcohols. When people talk about 'alcohol' in drinks, they really mean ethanol. All alcohols contain an —OH group attached to a carbon atom. The simplest alcohol is methanol, CH_3OH.

Alcohols form a homologous series like alkanes (see table). They have similar properties and their structures increase by units of CH_2.

CH_3OH methanol	
CH_3CH_2OH ethanol	
$CH_3CH_2CH_2OH$ propanol	
$CH_3CH_2CH_2CH_2OH$ butanol	

Table 1: The homologous series of alcohols

$$CH_3CH_2OH \qquad CH_3CH_3 \qquad HOH$$

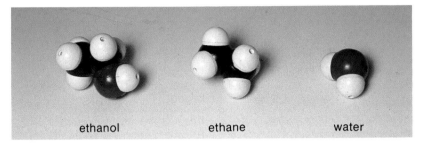

ethanol   ethane   water

Figure 1
The molecular models for ethanol, ethane and water

The properties of ethanol

Look at the structures of ethanol, ethane and water in figure 1. The structure of ethanol contains a C_2H_5 group like ethane and an —OH group like water. So we would expect ethanol to have some properties like ethane and some like water.

1 Ethanol is a colourless liquid (boiling point 78° C). It mixes with water in all proportions.

2 It is a very good solvent for molecular substances. Methylated spirits, widely used as a solvent in industry, contains 95% ethanol.

3 Ethanol is highly flammable. It burns very easily (like ethane) with a pale yellow flame forming carbon dioxide and water.

$$C_2H_6O + 3O_2 \rightarrow 2CO_2 + 3H_2O$$

Methylated spirits and ethanol are used as fuels in some countries.

4 Ethanol, unlike ethane or water, can be oxidized. The product is ethanoic acid (acetic acid).

ethanol   →   ethanoic acid

Wine goes sour after the bottle has been opened because the ethanol in it is oxidized to ethanoic acid. This is how wine vinegar is produced. Malt vinegar is made by a similar process using beer instead of wine.

Questions

1 Explain the words:
fermentation; enzyme; alcohol.
2 (a) Why is fermentation important?
(b) What are the main products of the fermentation of starch with yeast in anaerobic (no air) conditions?
(c) What causes the fermentation process?
3 The adverts say 'Don't drink and drive'. Why is this important?
4 (a) If an alcohol has *n* carbon atoms, how many hydrogen atoms will it have?
(b) Write a general formula for the family of alcohols.
5 Two substances have the molecular formula, C_2H_6O. Draw structural formulas for these two substances.

15 Energy from Fuels

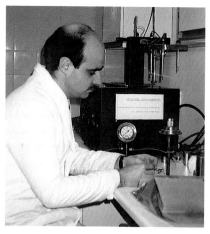

A food scientist using a bomb calorimeter

The amount of energy given out in a chemical reaction can be measured. Normally, we measure the heat given out or taken in when one mole of a substance reacts. This is called the **heat of reaction** and is given the symbol ΔH. For an exothermic reaction, the sign of ΔH is *negative*. This is because the chemicals have *lost* energy to the surroundings. For an endothermic reaction, ΔH is *positive* because the chemicals *gain* energy from the surroundings.

For example, when 1 mole (12 g) of carbon burns, 394 kJ of heat is given out. So we can write

$$C(s) + O_2(g) \rightarrow CO_2(g) \qquad \Delta H = -394 \text{ kJ}$$

It is useful to know how much energy is produced when a fuel burns. This can help us to decide the best fuel for a particular use. Figure 1 shows the apparatus we can use to measure the heat given out when a liquid fuel like meths burns. The heat produced warms the water in the metal can. If we measure the temperature rise of the water, we can work out the heat produced. We can find the mass of meths which is burnt from the loss in weight of the liquid burner. Then we can calculate the heat produced when one gram of the fuel burns. **Wear eye protection** if you try this experiment and remember that liquid fuels are highly flammable.

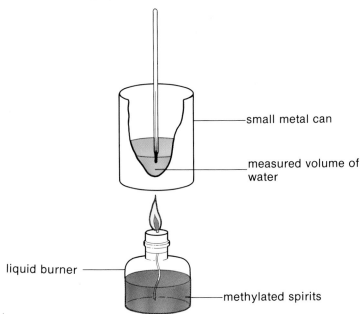

Figure 1

Mass of burner + meths at start of experiment	= 271.8 g
Mass of burner + meths at end of experiment	= 271.3 g
∴ Mass of meths burnt	= 0.5 g
Volume of water in metal can	= 250 cm³
∴ Mass of water in metal can	= 250 g
Rise in temp. of water	= 10° C

Table 1: The results of an experiment to measure the heat produced when meths is burnt

The results from one experiment are shown in table 1. In the experiment, 250 g of water are warmed up by 10° C.

We know that 4.2 joules of heat warm up 1 g of water by 1° C.

So, 250×4.2 J warm up 250 g of water by 1° C,

∴ $250 \times 4.2 \times 10$ J warm up 250 g of water by 10° C.

This amount of heat is produced by 0.5 g of meths

0.5 g meths produce $250 \times 4.2 \times 10$ J = 10 500 J
$$= 10.5 \text{ kJ}$$

∴ 1 g of meths produces 21 kJ

Meths is mainly ethanol (alcohol), C_2H_6O. If we assume that meths is pure ethanol ($M_r = 46$), then

the heat produced when 1 mole of ethanol burns = 46×21 kJ
$= 966$ kJ

So we can write

$$C_2H_6O(l) + 3O_2(g) \rightarrow 2CO_2(g) + 3H_2O(l) \quad \Delta H = -966 \text{ kJ}$$

Where does the energy come from?

In a reaction, bonds in the reactant chemicals are first broken. This requires energy. New bonds are then formed and this releases energy. If the energy released is greater than the energy required for the initial bond breaking, the reaction is exothermic.

Figure 2 shows the bond breaking, bond forming and energy changes when methane in natural gas reacts with oxygen.

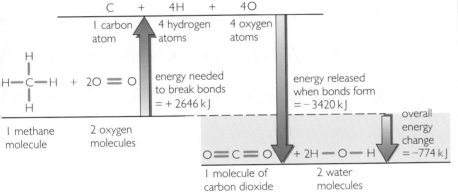

Figure 2

Notice in figure 2 that

- breaking chemical bonds is an *endothermic* process

- making chemical bonds is an *exothermic* process

The energy changes in chemical reactions, like those in figure 2, can be summarized in energy level diagrams. Figure 3 shows the energy level diagram for the reaction of methane with oxygen. This reaction is exothermic. Energy is lost during the reaction, so the products are at a lower energy level than the reactants.

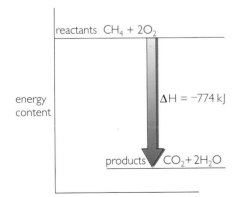

Figure 3

Questions

1 The simple apparatus in figure 1 does not give very accurate results. An accurate value for the heat produced when meths burns is about 30 kJ per gram.
(a) Why do you think the apparatus in figure 1 gives inaccurate results?
(b) How would you modify the apparatus in figure 1 to obtain more accurate results?
2 Draw a diagram of the apparatus you would use to find the heat produced when 1 g of a firelighter burns.
3 When 100 cm³ of 1.0 mol dm⁻³ hydrochloric acid is added to 100 cm³ of 1.0 mol dm⁻³ sodium hydroxide, the temperature of the mixture rises by 6.5° C.
(a) What mass of solution is warmed up? (Assume 1 cm³ of solution has a mass of 1 g.)
(b) How many joules warm up the mixture? (Assume 4.2 J warm up 1 g of solution by 1° C.)
(c) How many moles of hydrochloric acid react?
(d) How much heat is produced when 1 mole of hydrochloric acid reacts? This is called the **heat of neutralization** of hydrochloric acid.
(e) Write a balanced equation for the reaction involved in (d).
4 The energy change for the manufacture of lime (CaO) from limestone ($CaCO_3$) can be written as

$$CaCO_3(s) \rightarrow CaO(s) + CO_2(g)$$
$$\Delta H = +178 \text{ kJ}$$

(a) Is the reaction exothermic or endothermic?
(b) Do the products contain more or less energy than the reactants?
(c) Draw an energy level diagram for the reaction.

Section I: Activities

1   Plascup—new material for throw-away cups

What are the advantages and disadvantages of using throw-away food containers?

When was the last time you used a disposable (throw-away) cup? Did you burn your fingers? Did you enjoy the drink?

Suppose you are the manager of a firm which makes throw-away cups for hot or cold drinks. A new material called 'Plascup' has been produced with the following properties:

(i) It is easily coloured with dyes
(ii) It is a poor conductor of heat
(iii) It is easily cut with a knife
(iv) It starts to soften at 90° C
(v) It is biodegradable.

1 Which property or properties make 'Plascup' a good material for throw-away cups? Give a reason for your answer.
2 Which property or properties make 'Plascup' unsuitable for throw-away cups? Give a reason for your answer.
3 Suggest one other property of 'Plascup' that you might want to know before deciding to use it for throw-away cups.

2   Chemicals from crude oil—crossword

Across
1. (and 1 down) This method is used to split large molecules in the heavier fractions of crude oil (9, 8)
6. Electric currents are measured using this unit (3)
7. Liquid petroleum gas is sold for camping and caravanning under this trade name (3)
9. The formula of a compound produced when fuels do not burn fully (2)
11. Another name for the paraffin fraction from crude oil (8)
12. Symbol for a valuable metal used for jewellery (2)
13. This polymer was named after New York and London because it was first synthesized in both cities at the same time (5)
15. The simplest alkane (7)
18. Probably the most important fraction from crude oil (6)
19. All alkenes have names ending in this (3)

Down
1. (see 1 across)
2. Trichlorophenol is usually shortened to this (3)
3. An abbreviated name for liquid petroleum gas (3)
4. The formula of iodine monofluoride (2)
5. The common name for the most important synthetic plastic (9)
8. The metal used for galvanizing (4)
10. A small molecule which can add to itself (7)
14. The symbol for a noble gas used in advertising (2)
16. All alkanes have names ending in this (3)
17. The symbol of a gas used in weather balloons (2)

3 Windmills

Windmills have been used to drive machinery for hundreds of years. Recently, large windmills (sometimes called wind turbines) have been used to generate electricity. Giant windmills with a blade diameter of 50 metres have been built on the west coasts of Scotland and Wales. The photo on page 188 shows a giant wind turbine at Carmarthen Bay in Wales. Even so, one giant windmill cannot produce much electricity compared to a power station. One way to overcome this problem is to have windfarms with lots of giant windmills.

1 Make a list of the advantages of generating electricity using windmills in a windfarm rather than using a coal-fired power station.
2 Make a list of the disadvantages of generating electricity using windmills in a windfarm.

4 Margarine from vegetable oil

The apparatus below can be used to add hydrogen to vegetable oils and make margarine.

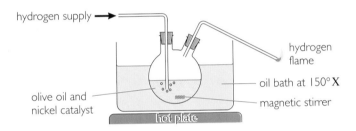

hydrogen supply →
hydrogen flame
oil bath at 150°X
olive oil and nickel catalyst
magnetic stirrer
hot plate

1 At the start of the experiment, hydrogen is passed through the apparatus for a few minutes before the flame is lit. Why is this done?
2 Choose a crude oil fraction from the table below for use in the oil bath. Explain your choice.

Fraction	Boiling range in °C
Gas	below 40
Petrol	40–160
Kerosine	160–250
Diesel oil	250–300
Lubricating oil	300–350

3 Give one reason for using a hotplate rather than a bunsen for heating.
4 Each molecule in olive oil contains three carbon-to-carbon double bonds. How many hydrogen atoms will have been added to one molecule when the reaction is complete?
5 What simple test would you use to show that the reaction had completely saturated the carbon-carbon double bonds in olive oil?
6 What has been done in the experiment to speed up the reaction?
7 Some brands of margarine are described as **polyunsaturated**. What does this mean?

Harnessing the energy of the wind is not a new idea. Windmills like this were used until the end of the last century to grind cereals to flour

Section I: Study Questions

1 The major source of the world's energy supply is fossil fuels. The chemical energy that they contain is converted to other more useful forms, e.g. heat and electrical energy. During the last thirty years there has been a considerable increase in demand for electrical energy and now other sources of energy are being investigated as a matter of urgency.

(a) Explain what is meant by the term 'fossil fuel'.

(b) Name *three* fossil fuels in use at the present time, each one existing in a different state of matter at room temperature and pressure.

(c) (i) Name the two elements that are present in the highest proportion in fossil fuels.

(ii) Write the symbol equations for the complete combustion of each of these elements in oxygen.

(d) Both of the reactions represented in part (c) are exothermic. Draw an energy-level diagram to show the energy changes that take place when one of the elements is burned in oxygen. Indicate clearly on the diagram the value for the heat of combustion of the element by labelling it $\triangle H$.

(e) Why are urgent efforts being made to find sources of energy as an alternative to the use of fossil fuels?

(f) State one source of energy, other than direct solar energy, which may be used as an alternative to fossil fuels.
NEAB

2 The following list shows some of the processes which are used in industry:
cracking
fermentation
polymerization
Using *one* example from industry in each case, describe briefly how each process may be used to manufacture a product useful in everyday life.

3 (a) The following diagram shows some of the reactions of ethene.

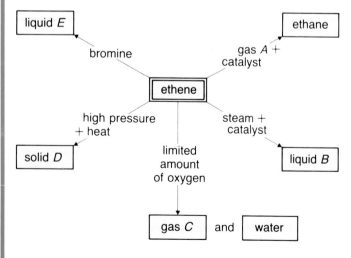

(i) Draw the structural formulas for ethane and ethene.

(ii) Give the names for:
gas A,   liquid B,   gas C,   solid D.

(iii) Write a symbol equation for the reaction between ethene and steam to form liquid B.

(iv) Give the name and draw the structural formula for liquid E.

(v) Ethene is an unsaturated compound. By means of gas A, ethene is converted into ethane, a saturated compound. Give one commercial application of this *type* of reaction.

(b) Liquid paraffin, which is a mixture of alkanes, is decomposed into products which include ethene. This process can be carried out in the apparatus below.

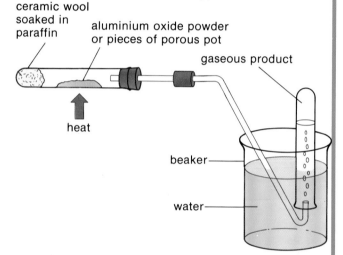

(i) What is the purpose of the pieces of porous pot?

(ii) What is the name given to this example of thermal decomposition?

(iii) Explain why changing alkanes into simpler alkanes and alkenes is important in the petrochemical industry.

4 Petrol is a mixture of liquids which are all hydrocarbons.

(a) Name the *two* elements that are found in hydrocarbon particles.

(b) When petrol vapour is exploded in a car engine, the elements mentioned in part (a) form exhaust gases. Give the name of *one* of these gases.

(c) Name *one* very toxic gas found in car exhaust fumes.

5 The substance ethanediol ($C_2H_6O_2$) is used as an antifreeze. It is made from ethene according to this equation:

$$C_2H_4 \xrightarrow{\text{water, air, catalyst}} C_2H_6O_2$$
$$\text{ethene} \qquad\qquad\qquad \text{ethanediol}$$

The relative atomic masses are C = 12; H = 1; O = 16.

(a) State *one* use of antifreeze.

(b) What is a catalyst?

(c) Calculate the relative molecular mass of ethene.

(d) Calculate the relative molecular mass of ethanediol.

(e) Calculate the mass of ethanediol that could be made from 5.6 g of ethene.

SECTION J
Reaction Rates

Explosives are used in mining to break up rock. Any explosion is an extremely fast reaction

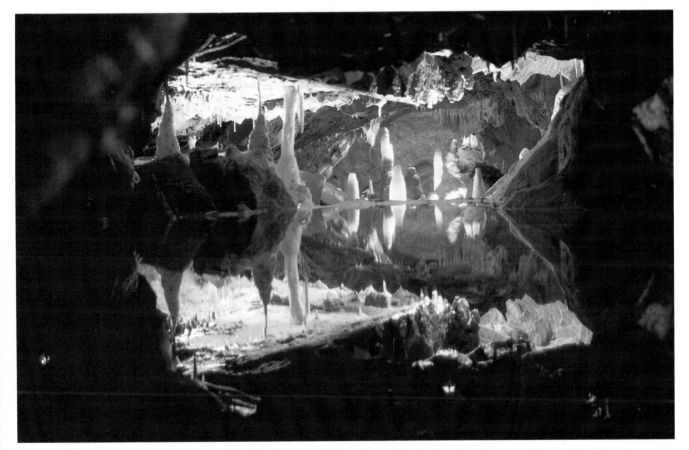

Stalagmites and stalactites in Gough's Cave, Cheddar, Somerset. Stalagmites and stalactites form by a very slow reaction. Limestone dissolves in rain water and is later precipitated again

1 How Fast?

Some reactions, like explosions, happen almost instantaneously

Limestone is weathered so slowly that it takes centuries before the effect is noticed. The face on this statue has been reacting with the carbonic acid in rainwater for centuries

The reactions which take place when food is cooked occur at a steady rate

Everyday we are concerned about how fast things happen. We want to know how quickly we can get to school, how fast a car is travelling or how soon the post will arrive. We are also interested in how fast some chemical reactions occur, how long we need to boil an egg or bake a cake and how quickly the rust appears on bikes and cars.

Different chemical reactions happen at different rates. Some reactions, like explosions, are so fast that they are almost instantaneous. A small explosion takes place during the 'pop' test for hydrogen. When a burning splint is put into a test tube containing hydrogen, the hydrogen reacts instantly with oxygen in the air forming water vapour.

$$2H_2(g) + O_2(g) \rightarrow 2H_2O(g)$$

Other reactions, like the rusting of steel and the weathering of limestone on buildings, happen so slowly that it may be years or even centuries before we notice their effects.

Most reactions take place at speeds somewhere between those described in the last two paragraphs. The reactions which take place when coal is burnt and when food is cooked are good examples of reactions which occur at steady rates.

Why are reaction rates important?

Most of the reactions in our bodies would never happen if the reacting substances were just mixed together. Fortunately each reaction in our bodies is helped along by its own special **catalyst**. Catalysts allow the chemicals to react more easily. The catalysts in living things are called **enzymes**. Without enzymes the reactions in your body would stop and you would die.

One of these enzymes is amylase. Amylase is present in saliva. It speeds up the first stage in the breakdown of starch in foods such as bread, potatoes and rice.

More and more industrial processes are being developed which use enzymes. The processes include brewing, baking and the manufacture of yoghurt, fruit juices and vitamins. In some of these processes, enzymes are extracted from living material such as plant extracts, animal tissues, yeast and fungi. Enzyme extracts of this kind are already used in cheese making, food processing and 'biological' washing powders.

Industrial chemists are not usually satisfied with just turning one substance into another. They want to carry out reactions faster and more cheaply. In industry, speeding up slow reactions makes them more economical because saving time usually saves money.

The key reaction in the manufacture of sulphuric acid is the contact process (section G, unit 1). This involves converting sulphur dioxide and oxygen to sulphur trioxide.

$$2SO_2 + O_2 \rightarrow 2SO_3$$

At room temperature, this reaction will not happen. But chemical engineers have found that the reaction takes place quickly at 450° C if a catalyst of vanadium(V) oxide or platinum is used. By using a catalyst, sulphuric acid can be made faster and more cheaply. This is important because sulphuric acid is a major industrial chemical.

A workman fitting a special catalyst section to the exhaust system of a car. The catalyst removes nitrogen oxide from the car's exhaust fumes

---

Questions

1 Make a list of the various ways in which gardeners speed up the growth rate of their plants.

2 Find out about the process of fermentation in the manufacture of beer and wine. What part do catalysts play in the process? Prepare a short report of your findings.

3 (a) State *three* ways in which food can be preserved and stored without 'going bad'.
(b) What conditions slow down the rate at which foods deteriorate (go bad)?

4 (a) How do you think a pressure cooker speeds up the rate at which food is cooked?
(b) Why do you think a microwave oven can speed up the rate at which food is cooked?

2 Studying Reaction Rates

A chemical reaction cannot happen unless particles in the reacting substances collide with each other.

This statement explains why reactions between gases and liquids usually happen faster than reactions involving solids. Particles in gases and liquids can mix and collide much more easily than particles in solids. In a solid, only the particles on the surface can react.

During a reaction, reactants are being used up and products are forming. The amount and the concentration of the reactants fall as the amount and the concentration of the products rise. So, we can measure reaction rates by measuring how much of a reactant is used up or how much of a product forms in a given time.

$$\therefore \text{Reaction rate} = \frac{\text{change in mass (or concentration) of a substance}}{\text{time taken}}$$

For example, when 0.1 g of magnesium was added to dilute hydrochloric acid, the magnesium reacted and disappeared in 10 seconds.

$$\therefore \text{Reaction rate} = \frac{\text{change in mass of magnesium}}{\text{time taken}}$$

$$= \frac{0.1}{10} \text{ g magnesium used up per second}$$

$$= 0.01 \text{ g s}^{-1}$$

Strictly speaking, this is the *average* reaction rate over the 10 seconds for all the magnesium to react. Although reaction rates are usually measured as changes in mass (or concentration) with time, we can also use changes in volume, pressure, colour and conductivity with time.

Why are gardeners advised to spread fertilizer pellets on their lawns just before rain is forecast?

This mechanic is tuning the car engine to adjust the rate at which petrol burns in the cylinders. The car's performance depends on the rate of this reaction

Time /minute	Mass of flask and contents /g	Decrease in mass /g	Decrease in mass for each minute interval /g
0	78.00	0	
1	76.50	1.50	1.50
2	75.50	2.50	1.00
3	74.95	3.05	0.55
4	74.60	3.40	0.35
5	74.41	3.59	0.19
6	74.33	3.67	0.08
7	74.30	3.70	0.03
8	74.30	3.70	0

Table 1: The results of one experiment to measure the rate of reaction between marble chips and dilute hydrochloric acid

Calculating reaction rates

The rate of reaction between small marble chips (calcium carbonate) and dilute hydrochloric acid was studied using the apparatus in figure 1. As the reaction occurs, carbon dioxide escapes from the flask and so the mass of the flask and its contents decrease.

$$CaCO_3(s) + 2HCl(aq) \rightarrow CaCl_2(aq) + H_2O(l) + CO_2(g)$$
marble chips

The cotton wool in the mouth of the flask stops liquid escaping from the flasks as the mixture fizzes. The results of one experiment are given in table 1. These results have been plotted on a graph in figure 2. During the first minute there is a decrease in mass of 1.5 g as carbon dioxide escapes.

∴ Average rate of reaction in the first minute

$$= \frac{\text{change in mass}}{\text{time taken}} = \frac{1.5}{1} = 1.5 \text{ g of carbon dioxide per minute}$$

(Notice that the units for reaction rate are grams per minute this time.) During the second minute (from time = 1 minute to time = 2 minutes), 1.0 g of carbon dioxide escapes.

∴ Average rate of reaction in the second minute

$$= \frac{1.0}{1} = 1.0 \text{ g of } CO_2 \text{ per minute.}$$

Notice that the reaction is fastest at the start of the reaction when the slope of the graph is steepest. During the reaction, the rate falls and the slope levels off. Eventually the reaction rate becomes zero and the graph becomes flat with a slope (gradient) of zero.

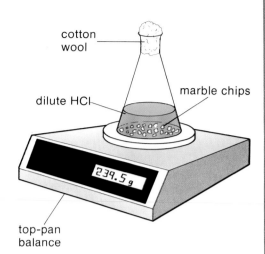

Figure 1
*Measuring the rate of reaction between marble chips and hydrochloric acid (**wear eye protection** if you try this experiment)*

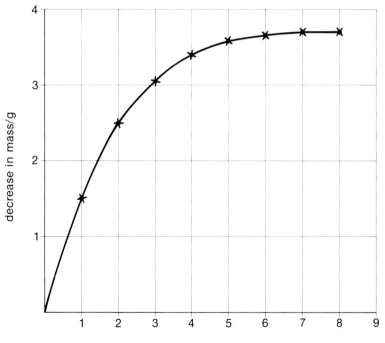

Figure 2   time of reaction/min

(y-axis: decrease in mass/g)

Questions

1 (a) When magnesium reacts with dilute hydrochloric acid, does the magnesium react faster at the start of the reaction or at the finish?
  (b) Give two reasons for your answer in part (a).

2 Look at the results in table 1 and the graph in figure 2.
  (a) What mass of carbon dioxide is lost from the flask in (i) the third minute (time 2 to 3 min); (ii) the fourth minute (time 3 to 4 min); (iii) the fifth minute (time 4 to 5 min)?
  (b) What is happening to the reaction rate as time passes?
  (c) Explain the change in reaction rate with time.
  (d) Why does the graph become horizontal after a while?

3 (a) Selective weedkillers can be added to a lawn either as solid pellets or as aqueous solutions. Which method will affect the weeds faster? Explain your answer.
  (b) The selective weedkiller 2,4D kills dandelions in a lawn, but not the grass. How do you think it works?

3 Making Reactions Go Faster

It's easier to get a fire started with sticks rather than logs

Anyone who has tried to light a fire knows that it is easier to burn sticks than logs. The main reason for this is that the sticks have a greater surface area. There is a larger area of contact with the air and so the sticks burn more easily.

> In general, reactions go faster when there is more surface area to react.

Surface area and reaction rate

The reaction between marble chips (calcium carbonate) and dilute hydrochloric acid can also be used to study the effect of surface area on reaction rate.

$$CaCO_3(s) + 2HCl(aq) \rightarrow CaCl_2(aq) + H_2O(l) + CO_2(g)$$

During the reaction, carbon dioxide escapes from the reacting mixture so there is a decrease in mass. The results are shown in figure 1. In experiment I, thirty *small* marble chips (with a total mass of 10 g) reacted with 100 cm^3 of dilute hydrochloric acid. In experiment II, six *large* marble chips (with a total mass of 10 g) reacted with 100 cm^3 of the same hydrochloric acid. There are more than enough marble chips in both experiments, so the acid will be used up first.

1 Why is the overall decrease in mass the same in both experiments?
2 Why do the graphs become flat?
3 Which graph shows the greater decrease in mass per minute at the start of the experiment?
4 Which experiment begins at the faster rate?
5 Why is the reaction rate different in the two experiments?

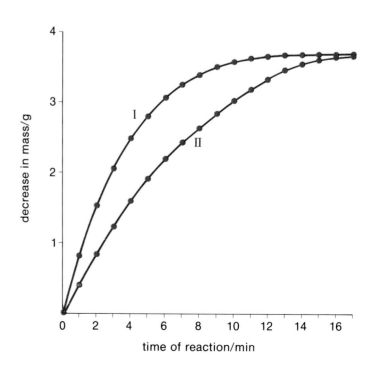

Figure 1

Concentration and reaction rate

Substances that burn in air burn much more rapidly in oxygen. Charcoal in a barbecue normally burns very slowly with a red glow. But, if you blow onto it so that it gets more air and more oxygen, it glows much brighter and may burst into flames. In oxy-acetylene burners, acetylene burns in pure oxygen. These burners produce such high temperatures that the flame will cut through sheets of metal.

$$C_2H_2 \quad + \quad 2\tfrac{1}{2}O_2 \rightarrow 2CO_2 + H_2O + \text{heat}$$
acetylene

> *Chemical reactions occur when particles of the reacting substances collide with each other.*

Collisions between acetylene molecules and oxygen molecules occur more often when oxygen is used instead of air. So, the reaction happens faster and gives off more heat when the concentration of oxygen is increased by using the pure gas. Pure oxygen can also be used to speed up chemical changes in the body. This can help the recovery of hospital patients, such as those suffering from extensive burns.

> *In general, reactions go faster when the concentration of reactants is increased.*

In reactions between gases, the concentration of each gas can be increased by increasing its pressure. Some industrial processes use very high pressures. For example, in the Haber process (unit 6 of this section), nitrogen and hydrogen are made to react at a reasonable rate by increasing the pressure to 250 times atmospheric pressure.

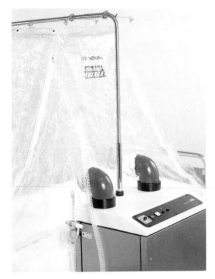

Pure oxygen is used in oxygen tents like this one, to speed up the recovery of hospital patients

Questions

1 It takes about 10 minutes to fry chips, but about 20 minutes to boil potatoes. Larger potatoes take even longer to boil.
  (a) Why do larger potatoes take longer to cook than small ones?
  (b) Why can chips be cooked faster than boiled potatoes?
  (c) Why can boiled potatoes be cooked faster in a pressure cooker?

2 Why do gaseous reactions go faster if the pressure of the reacting gases is increased?

3 Which of the following will affect the rate at which a candle burns?
the temperature of the air; the shape of the candle; the air pressure; the length of the wick.
Explain your answer.
State two other factors that will affect the rate at which a candle burns.

4 Design an experiment to investigate the effect of acid concentration on the weathering of limestone.
Do not carry out your experiment until it has been checked by your teacher.

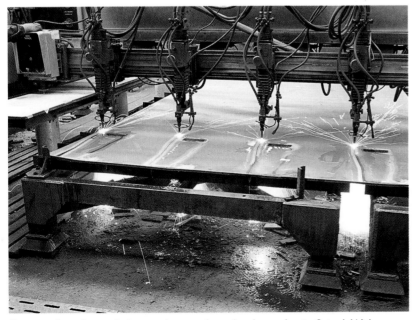

Oxy-acetylene flames being used to cut through a large sheet of steel. With oxy-acetylene burners, acetylene burns very rapidly in pure oxygen.

4 Temperature and Catalysts

Fish, meat and soft fruit can be kept for long periods in a deep freezer where the temperature is about −18° C. (The temperature in a house refrigerator is about 5° C)

Milk will keep for days in a cool refrigerator, but it turns sour very quickly if it is left in the sun. Other perishable foods, like fruit and cream, also go bad more quickly at higher temperatures. This is why we often keep them in a refrigerator.

Lots of other processes can be speeded up by increasing the temperature, or slowed down by reducing the temperature. The chemical reactions involved in baking and cooking would never happen unless the food was heated to a high temperature.

Why do chemical reactions go faster at higher temperatures?

At higher temperatures, particles move about faster. So they collide more often and this causes the reactions to go faster. But particles do not always react when they collide. Sometimes they collide too gently. The particles do not collide with enough energy for bonds to stretch and break. In some reactions, only the molecules with high energies can react. The same sort of thing happens in a car crash (figure 1).

> So there are two reasons why reactions go faster as the temperature rises:
> * The particles move faster and collide more often.
> * The particles collide with more energy, so more collisions result in a reaction.

Figure 1
If cars collide at very slow speeds (with low energy), they hardly dent each other. But if cars collide at high speeds (with high energy), they get smashed to pieces

Catalysts and reaction rates

Hydrogen peroxide solution ($H_2O_2(aq)$) decomposes very slowly into water and oxygen at room temperature. When manganese(IV) oxide is added, it decomposes very rapidly.

$$2H_2O_2(aq) \rightarrow 2H_2O(l) + O_2(g)$$

The manganese(IV) oxide helps the hydrogen peroxide to decompose, but it is not used up during the reaction. The manganese(IV) oxide left at the end weighs exactly the same as that at the start of the reaction. The manganese(IV) oxide has acted as a **catalyst**.

This reaction shows that:
Catalysts are substances which change the rate of chemical reactions without being used up during the reaction.

Liver and some plant tissues contain an enzyme called catalase which decomposes hydrogen peroxide in living things. As the photos show, liver can decompose hydrogen peroxide very rapidly

Most catalysts are used to speed up reactions, but a few can be used to slow reactions down. These substances are called negative catalysts or **inhibitors**. For example, glycerine is sometimes added to hydrogen peroxide as an inhibitor to slow down its rate of decomposition during storage. Hydrogen peroxide is used in industry to bleach textiles, paper and pulp.

Catalysts play an important part in the chemical industry. Sulphuric acid (section G, unit 9), petrol (section I, unit 8), margarine (section I, unit 9) and ammonia (section J, unit 7) are all produced by processes involving catalysts. The catalysts in many important industrial processes are transition metals or their compounds.

Catalysts allow substances to react more easily. They do this by helping bonds to break more easily. So the particles need less energy to react and the reaction is faster.

Biological washing powders and liquids contain enzymes which break down the chemicals in food, dirt and other stains on clothing. Enzymes are the catalysts in biological processes

Questions

1 The catalysts in many reactions are transition metals or their compounds. Make a list of these reactions. (The references on this page to other catalysed reactions will help you with this question.)

2 Discuss the following questions in groups of 2 or 3.

   (a) Why do plants grow faster in warm, wet weather than in cold, dry weather?

   (b) Why do gardeners add fertilizer to the soil?

   (c) Are fertilizers catalysts?

3 Describe an experiment that you could carry out to show that manganese(IV) oxide is not used up when it catalyses the decomposition of hydrogen peroxide.

4 The sketch graph below shows how the rates of decomposition of hydrogen peroxide (i) with a manganese dioxide catalyst and (ii) with an enzyme (catalase) catalyst are affected by changes in temperature. Explain the change in reaction rate with temperature in each case.

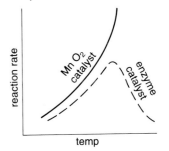

5 Reversible Reactions

Baking a cake is an irreversible reaction

Baking a cake, boiling an egg and burning natural gas are all one-way reactions. When a cake is baked or an egg is boiled, or natural gas burns, chemical reactions take place in the cake, the egg and the natural gas.

It is impossible to take the cake and turn it back into flour, sugar, water and fat. This also applies to the boiled egg and the carbon dioxide and water produced when natural gas burns.

$$CH_4(g) + 2O_2(g) \rightarrow CO_2(g) + 2H_2O(g)$$
methane in
natural gas

No matter what you do, carbon dioxide and water cannot be turned back into methane and oxygen. Reactions like this which cannot be reversed are called **irreversible reactions**. Most of the chemical reactions that we have studied so far are also irreversible, but there are some processes which can be reversed. For example, ice turns into water on heating,

$$H_2O(s) \xrightarrow{\text{heat}} H_2O(l),$$

but the ice reforms if water is cooled.

$$H_2O(l) \xrightarrow{\text{cool}} H_2O(s)$$

These two parts of this reversible process can be combined in one equation as:

$$H_2O(s) \underset{\text{cool}}{\overset{\text{heat}}{\rightleftarrows}} H_2O(l)$$

When blue hydrated copper sulphate is heated, it decomposes to white anhydrous copper sulphate and water vapour.

$$CuSO_4.5H_2O(s) \rightarrow CuSO_4(s) + 5H_2O(g)$$
blue $\qquad\qquad$ white

If water is now added, the change can be reversed and blue hydrated copper sulphate reforms.

$$CuSO_4(s) + 5H_2O(l) \rightarrow CuSO_4.5H_2O(s)$$
white $\qquad\qquad\qquad$ blue

These two processes can be combined in one equation as

$$CuSO_4.5H_2O(s) \underset{\text{mix reactants}}{\overset{\text{heat}}{\rightleftarrows}} CuSO_4(s) + 5H_2O(l)$$

Another important reversible reaction is that between nitrogen and hydrogen forming ammonia (figure 1).

Using the syringes, the mixture of hydrogen and nitrogen is pushed to and fro over the heated iron wool. The gases from the syringes are then ejected onto damp red litmus paper. The litmus paper turns blue showing that ammonia (the only common alkaline gas) has been produced.

$$N_2(g) + 3H_2(g) \rightarrow 2NH_3(g)$$
nitrogen $\quad$ hydrogen $\qquad\qquad$ ammonia
(hydrogen nitride)

> Reactions which can be reversed by changing the conditions or adding and removing reagents, are called **reversible reactions**.

Ice melts as it warms up in the drink, but if the drink is cooled the ice will reform. This is a reversible process

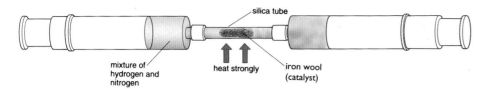

Figure 1
The reaction between nitrogen and hydrogen to
form ammonia

If the experiment is attempted without the iron wool, no ammonia is produced. If the experiment is repeated with different proportions of the reactants or at different temperatures then different amounts of ammonia are produced. These results show that:

> *the yields of products from reversible reactions depend on the conditions.*

Various manufacturing processes are based on reversible reactions. The manufacture of ammonia is, in fact, based on the reaction between nitrogen and hydrogen which we have just considered. This is taken further in the next unit.

In the manufacture of sulphuric acid, the contact process is also a reversible reaction.

$$2SO_2(g) + O_2(g) \rightleftarrows 2SO_3(g)$$

In this case, the reaction conditions involve a catalyst of vanadium(V) oxide at 450°C. These conditions ensure almost complete conversion of sulphur dioxide and oxygen to sulphur trioxide.

During a reversible reaction, the reactants are sometimes completely changed to the products. But, in other cases, the reactants are not *completely* converted to the products. For example, if ice and water are kept at 0°C, neither the ice nor the water seems to change. We say the two substances are in **equilibrium**. When two substances are in equilibrium like this, we replace the reversible arrows sign ($\rightleftarrows$) in the equation with the equilibrium arrows sign ($\rightleftharpoons$).

So, at 0°C, $H_2O(s) \rightleftharpoons H_2O(l)$

In the same way, nitrogen and hydrogen will come to equilibrium with ammonia in the apparatus shown in figure 1.

$$N_2(g) + 3H_2(g) \rightleftharpoons 2NH_3(g)$$

When equilibrium is reached, the concentrations of the reactants and products do not change any more. However, reactions are still going on both forwards and backwards.

This means that nitrogen and hydrogen are still reacting to form ammonia whilst ammonia is decomposing to re-form nitrogen and hydrogen. But these two processes, the forward and the backward reactions are taking place at the same speed, so there is no change in the overall amounts of any substance.

This is described as a **dynamic equilibrium** to indicate that substances are 'moving' (reacting) in both directions at equilibrium.

> **Caution:** *you should* **not** *attempt the experiment in figure 1 yourself.*

Questions

1 Explain the following terms: *irreversible reaction*; *reversible reaction*; *dynamic equilibrium*.

2 Look closely at figure 1 and the accompanying text.
(a) What conditions are chosen to speed up the reaction?
(b) How could you show that hydrogen is still present in the mixture of cases at the end of the reaction?
(c) Why should the Bunsen heating the silica tube be turned off before testing for any gases?

3 When purple hydrated cobalt chloride ($CoCl_2.6H_2O$) is heated, it changes to blue anhydrous cobalt chloride.
(a) Write an equation for this reaction.
(b) How is the reaction reversed?
(c) How is this reaction used as a test for water?

6 Reaction Rates and Industrial Processes

Fritz Haber (1868–1934). In 1904, Haber began to study the reaction between nitrogen and hydrogen. By 1908, he had found the conditions needed to make ammonia (NH$_3$). The ammonia could then be used to make fertilizers

Industrial chemists want to produce materials as fast and as cheaply as possible. In order to do this, they chose conditions which

1 increase the reaction rate and
2 use the most economic materials and methods.

One way to speed up a reaction is to increase the concentration of the reactants. At the same time, the products must be removed as fast as they form to prevent the reverse reaction happening.

Catalysts are also important in industrial processes. By using a suitable catalyst it is possible to carry out some processes that would otherwise be impossible. Other processes can be carried out at lower temperatures and lower pressures when a catalyst is used and this makes them more economical. Temperature and pressure are also chosen carefully in the manufacture of most chemicals.

The importance of these factors in industrial processes is well illustrated by the Haber process.

The Haber process

During the last century, the populations of Europe and America rose very rapidly. More crops and more food were needed to feed more and more people. So farmers began to use nitrogen compounds as fertilizers (unit 8 of this section). The main source of nitrogen compounds for fertilizers was sodium nitrate from Chile, but by 1900 supplies of this were running out.

Another supply of nitrogen had to be found or many people would starve. The obvious source of nitrogen was the air. But how could this unreactive gas be converted into ammonium salts and nitrates for use as fertilizers? The German chemist, Fritz Haber, solved the problem. In 1904, Haber began studying the reaction between nitrogen and hydrogen. By 1908 he had found the conditions needed to make ammonia (NH$_3$). Eventually, the Haber process became the most important method of manufacturing ammonia.

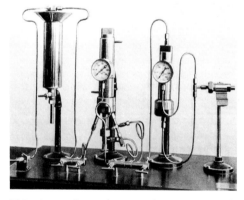

This picture shows the original apparatus used by Fritz Haber to make ammonia

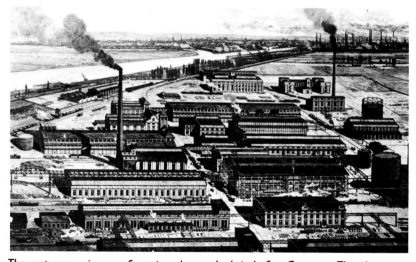

The vast ammonia manufacturing plant at Ludwigshafen, Germany. The plant was opened in 1913, only five years after Haber had found a cheap way of making ammonia

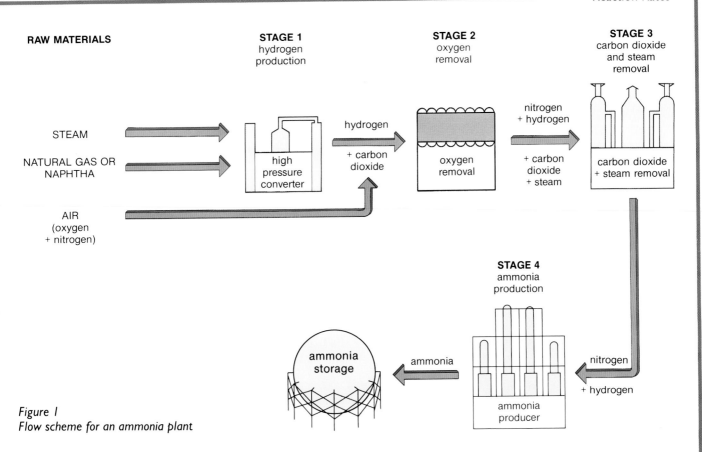

RAW MATERIALS

STEAM

NATURAL GAS OR
NAPHTHA

AIR
(oxygen
+ nitrogen)

STAGE 1
hydrogen
production

high
pressure
converter

hydrogen

+ carbon
dioxide

STAGE 2
oxygen
removal

oxygen
removal

nitrogen
+ hydrogen

+ carbon
dioxide
+ steam

STAGE 3
carbon dioxide
and steam
removal

carbon dioxide
+ steam removal

STAGE 4
ammonia
production

ammonia
producer

nitrogen

+ hydrogen

ammonia

ammonia
storage

Figure 1
Flow scheme for an ammonia plant

A flow scheme for the Haber process is shown in figure 1.

Stage 1 Steam reacts with natural gas (methane) or with naphtha from crude oil. This produces a mixture of hydrogen and carbon dioxide.

$$CH_4(g) + 2H_2O(g) \xrightarrow[+\ catalyst]{high\ temperature\ and\ pressure} CO_2(g) + 4H_2(g)$$

Stage 2 Air is added to the mixture. Oxygen in the air reacts with some of the hydrogen, forming steam.

$$2H_2(g) + O_2(g) \rightarrow 2H_2O(g)$$

The product gases now contain nitrogen, hydrogen, carbon dioxide and steam.

Stage 3 The carbon dioxide and steam are removed by passing the gases through concentrated potassium carbonate solution. This leaves a mixture of nitrogen and hydrogen.

Stage 4 This is the key reaction in the Haber process. Nitrogen and hydrogen are passed over

- a catalyst of iron

- at a pressure of 150–250 atmospheres and

- a temperature of 400° C

This converts 15–35% of the reactants to ammonia.

$$N_2(g) + 3H_2(g) \xrightarrow[400°C + iron\ catalyst]{150–250\ atm} 2NH_3(g)$$

The hot gases from the converter are cooled to liquefy the ammonia. The unreacted nitrogen and hydrogen are recycled.

Questions

1 What are the main factors that affect the rate of a chemical reaction?
2 (a) Name three industrial processes that use a catalyst. Say what the catalyst is in each case.
   (b) Why are catalysts important in industry?
3 At the beginning of this century, Haber synthesized ammonia from nitrogen and hydrogen. Why was this so important?
4 What conditions in the Haber process increase the rate of reaction between nitrogen and hydrogen?

7 Ammonia

Use	Approx. %
Fertilizers	75
Nitric acid	10
Nylon	5
Wood pulp and organic chemicals	10

The main uses of ammonia

Ammonia is an important chemical in industry and agriculture. Most of it is used for fertilizers and nitric acid (see the table). To understand these uses we must look at the properties of ammonia itself, which are listed in figure 1.

Ammonia is very soluble in water because it reacts with water to form a solution containing ammonium ions (NH_4^+) and hydroxide ions (OH^-).

$$NH_3(g) + H_2O(l) \rightleftharpoons NH_4^+(aq) + OH^-(aq)$$

The ammonia solution is a weak alkali. It is only partly dissociated into NH_4^+ and OH^- ions. The OH^- ions make the solution alkaline.

Ammonia as a base

Ammonia acts as a base in many reactions. Ammonia molecules pick up H^+ ions to form ammonium ions. So, ammonia reacts with acids to form ammonium salts. This is how fertilizers, such as ammonium nitrate ('Nitram') and ammonium sulphate, are made from ammonia.

$$\underset{\text{ammonia}}{NH_3(g)} + \underset{\text{nitric acid}}{HNO_3(aq)} \rightarrow \underset{\text{ammonium nitrate}}{NH_4NO_3(aq)}$$

Ammonia also reacts with hydrogen chloride gas to form a white smoke. The white smoke is tiny particles of solid ammonium chloride suspended in the air.

$$NH_3(g) + HCl(g) \rightarrow NH_4Cl(s)$$

Ammonia also acts as a base when it dissolves in water. In this case, NH_3 molecules take H^+ ions from the water to form NH_4^+ ions.

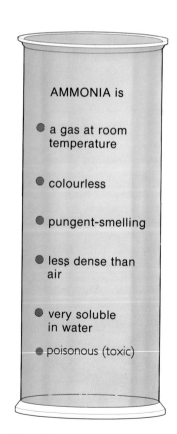

AMMONIA is

- a gas at room temperature

- colourless

- pungent-smelling

- less dense than air

- very soluble in water

- poisonous (toxic)

Figure 1
Properties of ammonia

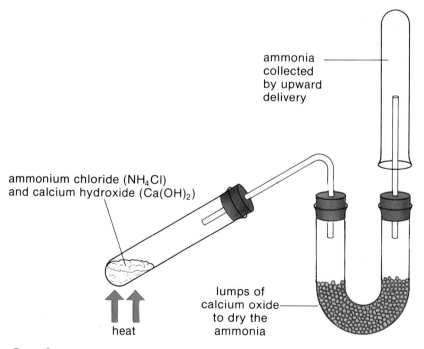

ammonia collected by upward delivery

ammonium chloride (NH_4Cl) and calcium hydroxide ($Ca(OH)_2$)

lumps of calcium oxide to dry the ammonia

heat

Figure 2
Making ammonia on a small scale. **Wear eye protection** and use a fume cupboard if you carry out this experiment

Making ammonia on a small scale

The easiest way to make a small amount of ammonia is to remove H^+ ions from NH_4^+ ions in an ammonium salt, like ammonium chloride (NH_4Cl). Alkalis containing OH^- ions will do this. Figure 2 shows how dry ammonia can be made by heating ammonium chloride and calcium hydroxide.

Ammonia to nitric acid—base to acid

About 10% of ammonia is used to manufacture nitric acid. Nitric acid is used to produce fertilizers such as potassium nitrate, and explosives like TNT (trinitrotoluene) and dynamite.

There are 3 stages in the manufacture of nitric acid.

1 Oxidizing the ammonia to nitrogen oxide (NO) using a platinum alloy catalyst at 900° C.

$$4NH_3 + 5O_2 \rightarrow 4NO + 6H_2O$$

2 Oxidizing the nitrogen oxide to nitrogen dioxide (NO_2) by mixing with air.

$$2NO + O_2 \rightarrow 2NO_2$$

3 Reacting the nitrogen dioxide and oxygen with water to form nitric acid (HNO_3).

$$4NO_2 + O_2 + 2H_2O \rightarrow 4HNO_3$$

The first two stages in this process can be carried out in a fume cupboard using the apparatus in figure 3. Ammonia evaporates from the solution and reacts with oxygen on the platinum spiral. Nitrogen oxide (NO) is produced which reacts with oxygen to form brown fumes of nitrogen dioxide in the flask.

Nitric acid is a typical mineral acid, like hydrochloric acid and sulphuric acid. The dilute acid shows typical acid reactions with indicators, metals, bases and carbonates. (See section G, unit 4.)

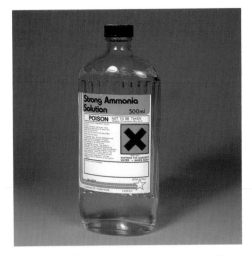

Ammonia solution can be used to clean toilets and sinks

Testing for ammonium ions, NH_4^+

If ammonium compounds are warmed with a little sodium hydroxide solution, ammonia is given off. This can be identified by its distinct smell (**care**) and by the fact that it turns damp red litmus paper blue.

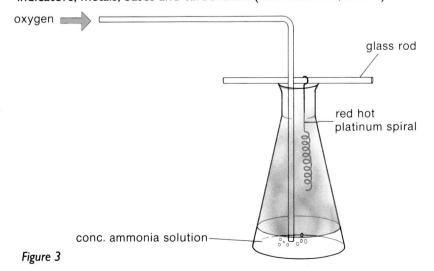

oxygen ➡

glass rod

red hot platinum spiral

conc. ammonia solution

Figure 3

Questions

1 Look at figure 2 and the small-scale preparation of ammonia.
   (a) Write an equation for the reaction involved.
   (b) Why is it necessary to dry the ammonia?
   (c) Why is concentrated H_2SO_4 *not* used to dry ammonia?
   (d) Why is ammonia *not* collected over water?
   (e) Why can ammonia be collected by upward delivery?
2 *True or false?*
Nitric acid
A forms salts called nitrites.
B forms soluble salts with all metals.
C is an important fuel.
D can be oxidized to ammonia.
3 Write equations for the reactions of nitric acid with (i) zinc; (ii) copper(II) oxide; (iii) potassium hydroxide; (iv) sodium carbonate.

8 Fertilizers

A bag of NPK fertilizer. The percentages of nitrogen, phosphorus oxides and potassium oxides are shown at the top of the bag

Fertilizer being bagged

Plants need essential elements (**nutrients**) to grow well. If crops are grown every year on the same land, these nutrients get used up. The soil becomes infertile and plants are stunted with poor seeds and small fruit.

The most important elements for plant growth are nitrogen, phosphorus and potassium. Plants need more of these three elements than other elements, so shortages of nitrogen, phosphorus and potassium are soon noticed. Table 1 shows the role of these three elements in plant growth and the effects of shortages. Any shortages of nitrogen, phosphorus or potassium must be replaced by adding fertilizers to the soil.

Fertilizers can be used as single compounds such as ammonium nitrate, or as mixtures of compounds containing nitrogen, phosphorus and potassium ('NPK' fertilizers). The proportions of nitrogen, phosphorus and potassium in NPK fertilizers are usually shown as % nitrogen (N), % phosphorus(V) oxide (P_2O_5) and % potassium oxide (K_2O).

Liquid ammonia being injected into the soil as a fertilizer. What are the advantages and disadvantages of using liquid ammonia?

Nutrient	Role in plant growth	Effect of shortage
Nitrogen	Essential for synthesis of proteins and chlorophyll	Plants are stunted, leaves become yellow (due to lack of chlorophyll)
Phosphorus	Essential for synthesis of nucleic acids (DNA)	Plants grow slowly. Small seeds and small fruit
Potassium	Plays a part in synthesis of carbohydrates and proteins	Leaves become yellow and curl inwards

Table 1: The role of nitrogen, phosphorus and potassium in plant growth and the effects of shortages of these elements

- **Nitrogen fertilizers** are usually nitrates or ammonium salts. Ammonium nitrate ('Nitram'), NH_4NO_3, is the most widely used fertilizer because it is soluble, it can be stored and transported as a solid and it has a high percentage of nitrogen (table 2). The higher the percentage of nitrogen the better, because less useless material needs to be stored and transported. Other nitrogen fertilizers are ammonium sulphate, urea and nitrochalk. Nitrochalk is a mixture of ammonium nitrate and chalk (calcium carbonate). This provides calcium for the soil as well as nitrogen and it also corrects soil acidity.

Fertilizer	Formula	Mass of one mole	Mass of nitrogen in one mole	% of nitrogen
Ammonium nitrate	NH_4NO_3	80 g	28 g	$\frac{28}{80} \times 100 = 35$
Ammonia	NH_3	17 g	14 g	$\frac{14}{17} \times 100 = 82$
Ammonium sulphate	$(NH_4)_2SO_4$	132 g	28 g	$\frac{28}{132} \times 100 = 21$
Urea	N_2H_4CO	60 g	28 g	$\frac{28}{60} \times 100 = 47$

Table 2: The percentage of nitrogen in different fertilizers

- **Phosphorus fertilizers** are manufactured mainly from phosphate rock containing calcium phosphate. This is insoluble in water, so it must be converted to soluble phosphorus compounds which plants can absorb through their roots. The phosphate rock is reacted with concentrated sulphuric acid. This converts it to 'super-phosphate'—a mixture of soluble calcium dihydrogenphosphate and insoluble calcium sulphate.

Although fertilizers are important in producing high yields of crops, problems are caused by their over use.

1 They may change the soil pH,
2 they may harm plants and animals in the soil,
3 they allow those elements not required by plants to accumulate in the soil,
4 they get washed out of the soil and lead to the pollution of rivers.

Fertilizers are essential for producing high crop yields year after year

Questions

1 (a) Why should a fertilizer be soluble?
(b) What are the problems in storing, transporting and using fertilizers?
(c) Make a list of the important properties of an ideal fertilizer.
2 Liquid ammonia has been used as a fertilizer in some countries.
(a) What are its advantages?
(b) What are its disadvantages?
3 (a) Why are fertilizers important?
(b) What problems are caused by their over use?
4 (a) Describe how you would make a sample of ammonium sulphate.
(b) Design an experiment to see if ammonium sulphate acts as a fertilizer for peas or beans.
5 (a) Why is NPK fertilizer so called?
(b) Rain water washes fertilizers into streams and rivers. What effect does this have on plants and animals that live in the water?
(c) Nitrochalk can act as a fertilizer and cure soil acidity. Explain why it can do both of these jobs.

9 The Nitrogen Cycle

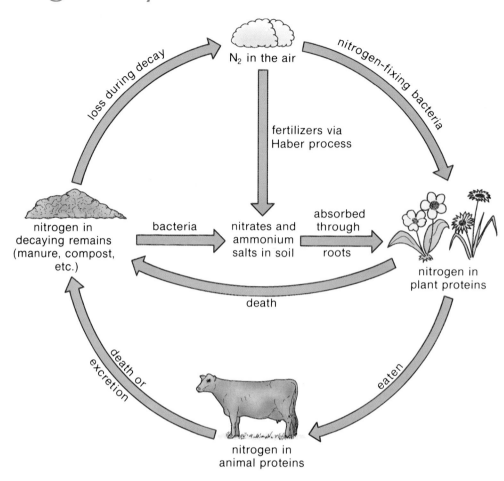

Figure 1 The nitrogen cycle

Some plants, called legumes, have nodules on their roots like these in the photo. These nodules contain bacteria which convert nitrogen from the air into compounds the plant can use for growth

In the last unit, we looked at man-made inorganic fertilizers like ammonium nitrate and super-phosphate. Large quantities of natural organic fertilizers are also used by farmers and gardeners. The most widely used organic fertilizers are manure and compost which contain decaying matter from animals and plants (see the lower half of figure 1). The nitrogen compounds in manure and compost are decomposed by bacteria to nitrates and ammonium salts. These substances dissolve in rain water and are absorbed through the roots of plants (centre of figure 1). The nitrates and ammonium salts are then used by plants to synthesize the chlorophyll and proteins that they need for growth. Animals have to eat plants and other animals in order to get the nitrogen and the proteins which they need.

Manure and compost are excellent fertilizers, but they take time to break down and there is not enough of them for all the crops we grow. Because of this, we need large amounts of inorganic fertilizers. Small quantities of the nitrogen that crops need are provided by nitrogen-fixing bacteria in the soil. These bacteria can convert nitrogen from the air into nitrogen compounds that can be used by plants. Some of these nitrogen-fixing bacteria live in nodules on the roots of plants such as peas, beans and clover (top right, figure 1). This way of converting atmospheric nitrogen to nitrogen compounds in plants is sometimes called *natural* nitrogen fixation. The conversion of atmospheric nitrogen to ammonia using the Haber process is described as *industrial* nitrogen fixation.

Although nitrogen is returned to the soil when living matter decays, some of the nitrogen compounds in the manure and compost are decomposed to nitrogen gas which escapes into the air (top left, figure 1). This is another reason why fertilizers are added to the soil in heavily cultivated areas.

The world food problem

There are about 4000 million people in the world. Nearly 3000 million of them are not properly fed and 400 million are starving. These figures suggest that we must increase food production, but the problem is not quite so simple.

- In Europe and North America, there are 'mountains' of surplus food but many poor countries cannot afford to buy it.
- In some poor countries, the rich have plenty to eat.
- Some countries need to export food to earn foreign currency even though some of their people are starving.

Three possible ways of reducing the food problem are as follows.

1 Increasing birth control. It is estimated that the world population will be 7000 million by the year 2000. Some people think that the population problem is more serious than the food problem.

2 Improving farming methods. This includes watering the deserts, preventing soil erosion, developing better varieties of crops, breeding better cattle and using improved pesticides.

3 Finding new food supplies such as farming the sea and growing bacteria on vegetable oils and cellulose to produce food.

Part of the European 'grain mountain'. Is it right that huge amounts of food are kept in store whilst thousands of people are starving?

During a lightning flash, nitrogen reacts with oxygen in the air to form nitrogen oxides. These react with rain water to produce nitric acid, which increases the nitrogen content of the soil

Questions

1 What is meant by (i) natural nitrogen fixation; (ii) industrial nitrogen fixation; (iii) the nitrogen cycle?

2 (a) What is the difference between man-made and natural-fertilizers?
(b) What happens to the nitrogen compounds in compost as it decays?
(c) Why do plants require nitrogen?
(d) Why do farmers and gardeners use man-made fertilizers as well as natural fertilizers?

3 What should the more developed nations do to solve the world food problem?

4 (a) What problems are caused by the over-use of fertilizers?
(b) Explain why these problems arise.

Section J: Activities

1 The Nitrogen Cycle Game

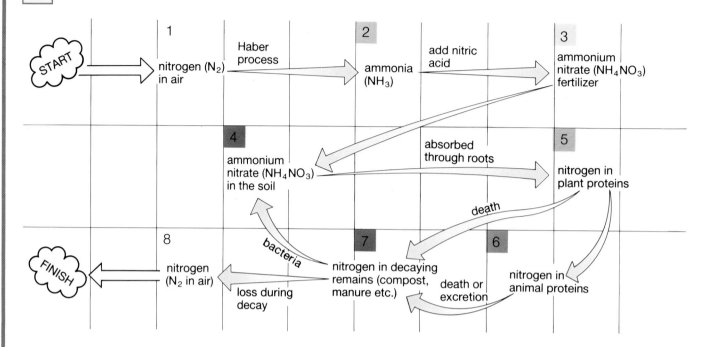

Instructions

A Fertilizer washed out of soil. Go back to 1.

B Fertilizer gets mixed with compost. Go on to 7.

C Fertilizer taken up by plants. Move to 5.

D Fertilizer returned to factory by farmer. Go back to 3.

E Fertilizer decomposes to ammonia. Go back to 2.

F Fertilizer accidentally eaten by cattle. Go on to 6.

G Soil is frozen by cold weather. Stay at 4.

H Important nutrients missing from soil. Stay at 4.

I Fertilizer changes the soil pH and plants cannot grow. Stay on 4 and miss a turn.

J Ideal conditions of pH and temperature for plant growth. Move to 7.

The Nitrogen Cycle Game is a mixture of 'Monopoly' and 'Trivial Pursuits'. It is based on fertilizers and the nitrogen cycle.

Chance cards

First, make 10 CHANCE cards by writing the letters A to J on 10 pieces of card. Shuffle these cards and place the pile face downwards on the table.

Rules

(i) Any number of people can play the game.

(ii) You can play as individuals or in teams.

(iii) Each person (or team) starts a number 1 in the diagram and moves to the next number if they give a correct answer to a question from the other team.

(iv) If you answer a question wrongly, you do *not* move.

(v) When a team are asking questions, they should make up questions using pages 228 to 235.

(vi) Questions are put to each person (team) in turn.

(vii) At number 4, each person (team) takes a CHANCE card. The letter on the CHANCE card tells you how you must move according to the instructions on the left.

(viii) The winning person (team) is the first to reach number 8.

2 AMMChem plant proposed for Oakbridge

AMMChem plant for Oakbridge

Yesterday, Sandra Bigg, Mayor of Oakbridge, admitted that Council officials had met representatives of AMMChem. AMMChem are the world's largest manufacturer of fertilisers. Mayor Bigg said. "The firm are seeking permission to build a chemical plant on the North side of Oakbridge."

AMMChem produces ammonia and nitric acid and uses them for the manufacture of ammonium nitrate (Nitram). It also sells large quantities of ammonia and nitric acid to other companies.

Mrs. Bigg welcomed the plant. She said, "It will help the town's economy. The plant will employ 150 local people. It is also estimated that each person employed by AMMChem will create jobs for another 3 or 4 people in local businesses. This will reduce unemployment in the town." At present, about 2000 people in Oakbridge are unemployed. This is 10% of the workforce.

Mayor Bigg also pointed to two other benefits from the plant. AMMChem would pay business rates to Oakbridge Council at the commercial level. This would make a large increase in revenue and hold down the increases in council tax to home owners. Farmers near Oakbridge would save about £10 per tonne on Nitram fertiliser because they would be paying lower transport costs. A representative for the National Farmer's Union has estimated that about 500 tonnes of Nitram fertiliser are used each year on farms near Oakbridge.

Herb Green, Secretary of the Oakbridge Environmental Group, was asked for his views. He said, 'We are concerned about emissions of ammonia and nitrogen oxides from the plant and possible accidents when chemicals are being transported. How will these affect the beautiful countryside around Oakbridge? Just imagine what would happen if the plant was not working properly and excessive amounts of ammonia or nitric acid got into the River Soak? Why can't the Mayor persuade cleaner industries to move into the area? Anyway, farmers are using less artificial fertilisers these days and growing more organic food."

Earlier today, Dr. Ken Bond, Managing Director of AMMChem issued the following statement:

"Some chemical companies have caused environmental problems but AMMChem have an excellent record. Our Safety Division checks processes and equipment continuously and ensures that all our operations are safe and pollution free.

No industry is without risks. Ammonia is very toxic and it is manufactured at high pressure and high temperature. An accident at the plant or during transportation would create a serious health hazard. But, accidents are very rare. AMMChem have been involved in only one death from such accidents in the last ten years. In recent years, industry and society have begun to realize their joint responsibility for the production of chemicals which will benefit society."

Suppose you live in Oakbridge.
l Make a list of the possible problems from the AMMChem plant.
l Make a list of the possible benefits from the AMMChem plant.
l Write a letter to the Oakbridge Chronicle expressing your views about the proposed plant. (You may want to support the proposal or to be very critical of it.)
l Suppose you are a town councillor in Oakbridge. Make a list of other points you would want to consider or other information which you would need before deciding whether AMMChem should build the proposed plant.

Section J: Study Questions

1 When manganese(IV) oxide is added to hydrogen peroxide, oxygen and water are produced. The oxygen can be collected and measured at timed intervals. The equation for the reaction is

$$2H_2O_2 \rightarrow 2H_2O + O_2$$

The results for one experiment are shown below. 1.0 g of manganese(IV) oxide was added to 100 cm³ of a solution of hydrogen peroxide.

Time/minutes	0	1	2	3	4	5	6	7	8
Volume/cm³		1	20	33	44	52	58	59	60	60

(a) (i) Plot a graph of volume of oxygen (vertically) against time. Label this curve X.
(ii) Mark on curve X the time at which the rate of reaction is fastest. Label this point Y. Why have you chosen this point?
(iii) The experiment is repeated using 1.0 g of the same catalyst, 50 cm³ of the hydrogen peroxide solution and 50 cm³ of water. Sketch the curve expected from this experiment between the same axes. Label this curve Z.
(b) The speed of a chemical reaction depends on:

• the rate at which the particles collide,
• the energy possessed by these particles.

Use these factors to explain each of the following:
(i) heating the solution of hydrogen peroxide increases the speed of the reaction;
(ii) 1.0 g of a finely powdered catalyst gives a faster reaction than 1.0 g of small lumps of the same catalyst.

2 This graph shows the total volume of hydrogen produced in the reaction of magnesium ribbon with excess dilute hydrochloric acid over a period of time.

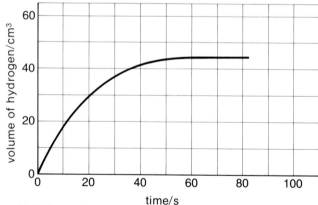

(a) What volume of hydrogen has been produced after 15 s?
(b) How long does it take to produce 28 cm³ of hydrogen?
(c) Use the graph to work out the volume of hydrogen produced after 100 s.
(d) Sketch *on the graph* the results that you would expect to obtain if the same mass of magnesium was treated with more concentrated acid.

SEG

3 Farmers use large amounts of nitrogen-containing fertilizers. Some of these fertilizers are washed off the farm land by rain into streams and lakes. Also present in streams and lakes are phosphates from domestic detergents. The dissolved fertilizers and phosphates increase the amount of chemicals needed by plants in the water. The surface of the water then becomes covered with algae. Because of this, the plants below the surface die. When the algae decays, the amount of dissolved oxygen in the water is lowered.
(a) Give the chemical name of a fertilizer which contains nitrogen.
(b) Why are fertilizers used on farms?
(c) Name an element, other than nitrogen, which is needed for plant growth.
(d) How do phosphates get into the water system?
(e) What would be the effect of lowering the amount of dissolved oxygen in streams and lakes? **MEG**

4 Ammonia is an important chemical in the manufacture of fertilizers. The Haber process is the industrial method by which ammonia is made from nitrogen and hydrogen.

The product from the Haber process consists of ammonia together with unreacted nitrogen and hydrogen. The percentage of ammonia in this mixture varies with temperature and pressure as shown in the table below.

Temp/° C	Percentage of NH₃ in mixture at pressure of:		
	1 atm	100 atm	1000 atm
200	15.3	80.6	98.3
300	2.2	52.1	92.6
400	0.44	25.1	79.8
500	0.15	10.4	57.5
600	0.05	4.5	31.4
700	0.02	2.1	12.9
800	0.01	1.2	—

(a) Write a balanced equation for the reaction involved in the Haber process.
(b) Use the table to comment on the effects of changing the temperature and the pressure on the percentage of ammonia produced.
(c) (i) From the table, state the conditions that produce the largest percentage of ammonia.
(ii) What conditions are normally used in the Haber process?
(iii) Why do you think that the conditions normally used in the Haber process are not the same as those conditions which give the largest percentage of ammonia.
(d) (i) What is the effect of an iron catalyst on the Haber process?
(ii) Explain the effect of the catalyst.
(e) The boiling points of the chemicals in the mixture produced by the Haber process are:

Ammonia   −34° C
Nitrogen   −196° C
Hydrogen   −253° C

Using this information, describe how ammonia can be separated from the mixture.

Atomic Structure and Radioactivity

Scientists working in radiation laboratories must handle radioactive isotopes by remote control from the other side of very thick glass windows

1 Inside Atoms

J.J. Thomson, Professor of Experimental Physics at Cambridge University and winner of the Nobel Prize for Physics in 1906

Less than a century ago scientists believed that atoms were solid particles like tiny snooker balls. Since then, experiments have shown that all atoms are made of three types of particles—protons, neutrons and electrons. In this unit we shall study some of the evidence for these particles.

1897: Thomson discovers electrons

In 1897, J.J. Thomson was investigating the way that gases conduct electricity. When he applied 15 000 volts across the electrodes of a tube containing a gas, the glass walls glowed a bright green colour. Rays travelling in straight lines from the cathode hit the glass and made it glow. Thomson called these rays **cathode rays** because they come from the cathode. Experiments with a narrow beam of cathode rays (figure 1) showed that they could be deflected by an electric field. When cathode rays passed between charged plates, they always bent towards the positive plate. This showed that the rays were negatively charged.

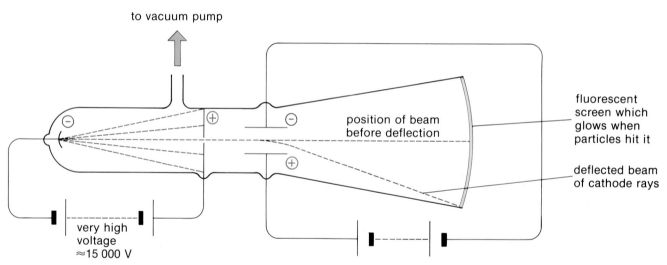

Figure 1
Deflection of cathode rays by an electric field

Further study of the deflection showed that cathode rays consisted of negative particles 1840 times lighter than hydrogen atoms. Thomson called these tiny negative particles **electrons**. The cathode rays were always the same, no matter what gas was present in the tube or what the electrodes were made of. This suggested that the electrons in all substances were identical.

1909: Geiger and Marsden explore the nucleus

Since atoms are neutral, they must contain positive charge to balance the negative charge on their electrons. Geiger and Marsden found a method of probing inside atoms using alpha particles from radioactive substances as 'bullets'. Alpha particles are helium ions, He^{2+}. When alpha particles from radium were fired at thin sheets of metal foil, most of the alpha particles passed straight through the foil. But, some of the alpha particles were deflected by the foil and a few of them even appeared to bounce back from it (figure 2).

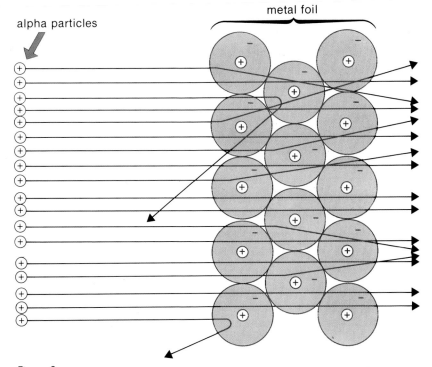

Ernest Rutherford who succeeded J.J. Thomson as Professor of Experimental Physics at Cambridge University in 1919. Rutherford received the Nobel Prize for Chemistry in 1908 while he was working at Manchester University

Figure 2
Most alpha-particles pass straight through the foil, some are deflected, but a few rebound from the foil

1911: Rutherford explains the structure of atoms

Rutherford explained Geiger and Marsden's results by suggesting that atoms in the foil must have a very small positive **nucleus**. Surrounding this is a much larger region of empty space in which the electrons move. Most of the positive alpha particles pass straight through the large empty space where the electrons are moving. A few alpha particles pass close to the positive nucleus and get deflected. Occasionally an alpha particle approaches a nucleus head-on. When this happens, the positive alpha particle is repelled by the positive nucleus and bounces back. Rutherford suggested that the structure of an atom could be compared to a miniature solar system. Each atom has a positive nucleus, orbited by tiny negative electrons like planets orbiting the Sun. He suggested that the positive charge of the nucleus was provided by positive particles which he called **protons**. Hydrogen, the smallest atom, has one proton in the nucleus, balanced by one orbiting electron. Atoms of helium, the next smallest, contain two protons and two electrons; lithium atoms have three protons and three electrons, and so on.

Questions

1 What are (i) cathode rays; (ii) electrons; (iii) protons?
2 (a) What evidence is there that electrons are (i) negatively charged; (ii) the same in all substances?
   (b) What sort of experiments might be carried out to show that electrons are about 2000 times lighter than hydrogen atoms?
3 Why did Geiger and Marsden's experiment suggest that atoms have a small positive nucleus surrounded by a much larger region of empty space?
4 How is the position of an element in the periodic table related to the number of electrons in its atoms?
5 When electrons pass between charged plates they are deflected towards the positive plate. What will happen when alpha particles pass between charged plates?

2 The Structure of Atoms

	Hydrogen atom	Helium atom
Number of protons	1	2
Number of neutrons	0	2
Relative mass	1	4
Relative atomic mass	1	4

Table 1: the relative atomic masses of hydrogen and helium

In spite of Rutherford's success in explaining atomic structure, one big problem remained. Hydrogen atoms contain one proton and helium atoms contain two protons. So the relative atomic mass of helium should be two, since the relative atomic mass of hydrogen is one. Unfortunately, the relative atomic mass of helium is four and *not* two.

James Chadwick, one of Rutherford's colleagues, showed where the extra mass in helium came from. Chadwick discovered that the nuclei of atoms contained *uncharged* particles as well as positively charged protons. Chadwick called these uncharged particles **neutrons**. Further experiments showed that neutrons have the same mass as protons, so Chadwick was able to explain the problem concerning the relative atomic masses of hydrogen and helium (table 1). Hydrogen atoms have one proton, no neutrons and one electron. Since the mass of the electron is almost zero compared to the proton and neutron, a hydrogen atom has a relative mass of one unit. Helium atoms have two protons, two neutrons and two electrons. The two protons and two neutrons give a helium atom a relative mass of four units. Thus, a helium atom is four times as heavy as a hydrogen atom and the relative atomic mass of helium is four.

James Chadwick discovered neutrons in 1932 when he was working with Rutherford in Cambridge. In 1935, he won a Nobel Prize for this achievement and soon afterwards he became Professor of Physics at Liverpool University

Protons, neutrons and electrons

We now know that all atoms are made up from three basic particles—protons, neutrons and electrons. The nuclei of atoms contain protons and neutrons. Both of these particles have a mass about the same as a hydrogen atom. Neutrons have no charge, but protons have a positive charge. Moving around the nucleus are electrons that are negatively charged. The charge on one electron is opposite in sign but equal in size to the charge on one proton. So, the negative charge on an electron is just balanced by the positive charge on one proton. In an atom, the electrons are arranged in layers or **shells** at different distances from the nucleus. The mass of the electron is so small that it can be ignored when working out the total mass of the atom. The positions, masses and charges of these three sub-atomic particles are shown in table 2.

Particle	Position	Mass (relative to a proton)	Charge (relative to that on a proton)
Proton	Nucleus	1	+1
Neutron	Nucleus	1	0
Electron	Shells	$\frac{1}{1840}$	−1

Table 2: properties of the three sub-atomic particles

Different atoms have different numbers of protons, neutrons and electrons. The hydrogen atom is the simplest of all atoms. It has one proton in the nucleus, no neutrons and one electron (figure 1). The next simplest atom is that of helium, with two protons, two neutrons and two electrons. The next, lithium, has three protons, four neutrons and three electrons. Some of the heavier atoms can have large numbers of protons, neutrons and electrons. For example, atoms of uranium have 92 protons, 92 electrons and 143 neutrons. Notice that hydrogen, the simplest atom and the first element in the periodic table, has one proton. Helium, the second element in the periodic table, has two protons. Lithium, the third element in the periodic table, has three protons and so on. Thus, the position of an element in the periodic table tells you how many protons it will have. Furthermore, an atom must always have an equal number of protons and electrons, so that the positive charges (on the protons) balance the negative charges (on the electrons).

If the nucleus of an atom was enlarged to the size of a pea and put on top of Nelson's Column, the electrons furthest away would be on the pavement

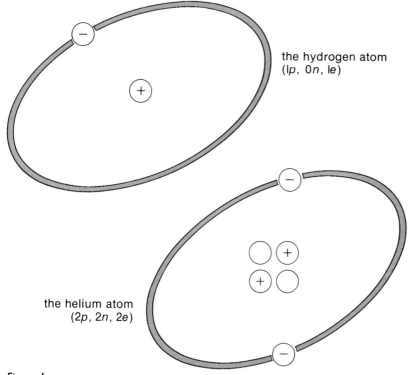

the hydrogen atom
(1p, 0n, 1e)

the helium atom
(2p, 2n, 2e)

Figure 1
Protons, neutrons and electrons in a hydrogen atom and a helium atom
($\oplus$ ≡ proton, $\bigcirc$ ≡ neutron, $\ominus$ ≡ electron)

Questions

1 What are the charges, relative masses and positions in an atom of protons, neutrons and electrons?
2 How many protons, neutrons and electrons are there in one (i) H atom; (ii) H^+ ion; (iii) Li atom, (iv) Li^+ ion?
3 Oxygen is the eighth element in the periodic table. How many protons and electrons are there in one (i) O atom, (ii) O^{2-} ion, (iii) O_2 molecule, (iv) H_2O molecule?
4 Lithium atoms have three protons, four neutrons and three electrons. Why is the relative atomic mass of lithium about seven?
5 (a) Make a list of important dates, scientists and facts in the development of our ideas about atomic structure.
  (b) How have ideas and theories about atomic structure changed since about 1890?

3 Atomic Number and Mass Number

Only hydrogen atoms have one proton. Only helium atoms have two protons. Only lithium atoms have three protons, and so on. This shows that the number of protons in an atom decides which element it is. Because of this, scientists have a special name for the number of protons in the nucleus of an atom. They call it the **atomic number** or proton number (symbol Z). Thus, hydrogen has an atomic number of one (Z = 1), helium has an atomic number of two (Z = 2), lithium has an atomic number of three (Z = 3) and so on. Aluminium, the thirteenth element in the periodic table with 13 protons and 13 electrons, has an atomic number of 13.

Protons alone do not account for all the mass of an atom. Neutrons in the nucleus also contribute to the mass. Therefore, the *mass* of an atom depends on the number of protons and the number of neutrons added together. This number is called the **mass number** or nuclear number of the atom (symbol A). So,

> *Atomic number = number of protons.*
> *Mass number   = number of protons + number of neutrons.*

Thus, hydrogen atoms (with one proton and no neutrons) have a mass number of one (A = 1). Helium atoms (two protons and two neutrons) have a mass number of four (A = 4) and lithium atoms (three protons and four neutrons) have a mass number of seven (A = 7). We can write the symbol $^{7}_{3}Li$ (figure 1) to show the mass number and the atomic number of a lithium atom. The mass number is written at the *top* and to the left of the symbol. The atomic number is written at the *bottom* and to the left. A lithium ion is written as $^{7}_{3}Li^{+}$. A sodium atom (11 protons and 12 neutrons) is written as $^{23}_{11}Na$. An electron (mass almost zero, charge −1) can be written as $^{0}_{-1}e$.

This photograph shows evidence for the two isotopes in neon, neon-20 and neon-22. Notice that the trace from neon-20 is much more prominent than that from neon-22. What does this tell you about the two isotopes? What does CO represent?

Figure 1

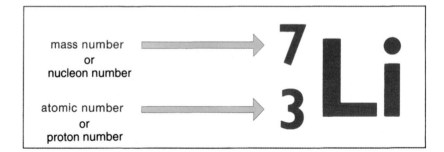

Using the periodic table (section E), we can predict the atomic number of an element, because the elements are arranged in order of atomic number. Therefore, the sixth element in the periodic table has an atomic number of six; the twentieth element an atomic number of 20 and so on.

Isotopes

Many elements have relative atomic masses which are nearly whole numbers. For example, the relative atomic mass of nitrogen is 14.007 and that of sodium is 22.99. This is not surprising, since the mass of an atom depends on the mass of protons and neutrons in its nucleus and the relative mass of both these particles is 1.00.

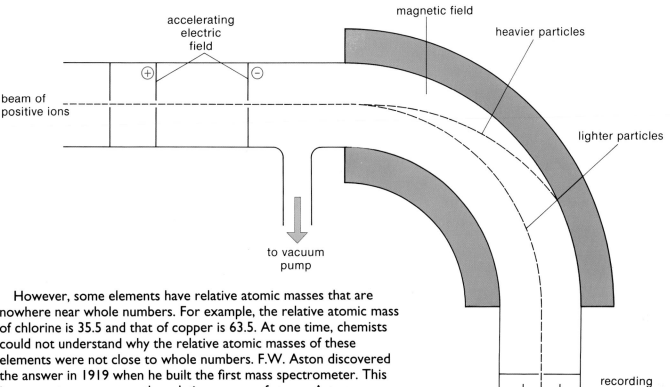

accelerating
electric
field

magnetic field

heavier particles

beam of
positive ions

lighter particles

to vacuum
pump

Figure 2
*A mass spectrometer. A beam of positive ions is
accelerated by an electric field and then
deflected by a magnetic field. The amount of
deflection depends on the mass of the particles
and the strength of the magnetic field. Lighter
particles are deflected more than heavier
particles. As the magnetic field is increased in
strength, the heavier particles are deflected
more and measured on the recording
instrument*

recording
instrument

However, some elements have relative atomic masses that are
nowhere near whole numbers. For example, the relative atomic mass
of chlorine is 35.5 and that of copper is 63.5. At one time, chemists
could not understand why the relative atomic masses of these
elements were not close to whole numbers. F.W. Aston discovered
the answer in 1919 when he built the first mass spectrometer. This
instrument can compare the relative masses of atoms. Aston
discovered that some elements contained atoms with different
masses. When atoms of these elements were ionized and passed
through a mass spectrometer, the beam of ions separated into two
or more paths (figure 2). This suggested that one element could have
atoms with different masses. These *atoms of the same element with
different masses are called* **isotopes**. Each isotope has a relative mass
close to a whole number, but the average atomic mass for the
mixture of isotopes is not always close to a whole number. This is
studied further in the next section.

*Aston's original mass spectrometer. Positive ions were accelerated along the
cylindrical metal tube at the top of the instrument. The coils of the electromagnet
produced a magnetic field which deflected the particles*

Questions

1 Explain the following: (i) *atomic
number;* (ii) *mass number;* (iii) *isotope.*
2 (a) What is the atomic number of
fluorine?
(b) How many protons, neutrons
and electrons are there in one
fluorine atom of mass number 19?
3 (a) What do 16, 8, 2− and O
mean in the symbol, $^{16}_{8}O^{2-}$?
(b) How many protons, neutrons
and electrons are there in one
$^{23}_{11}Na^+$ ion?
4 Why do some elements have
relative atomic masses which are not
close to whole numbers?

4 Isotopes

Isotopes have the same
Number of protons
Number of electrons
Atomic number
Chemical properties

Isotopes have different
Numbers of neutrons
Mass numbers
Physical properties

Table 1: The similarities and differences between isotopes of the same element

Isotopes are atoms of the same element with different masses. All the isotopes of one element have the same number of protons. Therefore, they have the same atomic number. Since isotopes have the same number of protons, they must also have the same number of electrons. This gives them the same chemical properties because chemical properties depend upon the number of electrons in an atom and the way in which these electrons are transferred and shared during reactions.

Isotopes do, however, contain different numbers of neutrons. This means **they have the same atomic number but different mass numbers**. For example, neon has two isotopes. Each isotope has 10 protons and 10 electrons and therefore an atomic number of 10. But one of these isotopes has 10 neutrons and the other has 12 neutrons. Their mass numbers are therefore 20 and 22 (figure 1). They are sometimes called neon-20 and neon-22. These 2 isotopes of neon have the same chemical properties because they have the same number of electrons, but they have different physical properties because they have different masses. Samples of $^{20}_{10}Ne$ and $^{22}_{10}Ne$ have different densities, different melting points and different boiling points. The similarities and differences between isotopes of the same element are summarized in table 1.

Figure 1
The two isotopes of neon

	neon-20	neon-22
	$^{20}_{10}Ne$	$^{22}_{10}Ne$
number of protons	10	10
number of electrons	10	10
atomic number	10	10
number of neutrons	10	12
mass number	20	22

Uranium ore being treated in giant tanks at the Rossing Mine in Namibia

Chemists can obtain samples of uranium with a higher percentage of uranium-235 because of the different physical properties of uranium-235 and uranium-238. Uranium-235 is needed for use in nuclear reactors. Natural uranium contains only about 0.7% of uranium-235 and 99.3% of uranium-238. Nuclear reactors use uranium with 3 or 4% of uranium-235. The natural uranium is converted to uranium hexafluoride (UF_6) which is very volatile. The uranium hexafluoride is vaporized and allowed to diffuse through a series of porous barriers. Particles of $^{235}UF_6$ are slightly lighter than those of $^{238}UF_6$. So they can move faster and diffuse faster than those of $^{238}UF_6$. After repeated diffusion through the porous barriers, the uranium hexafluoride contains 3 or 4% $^{235}UF_6$. Finally, this can be converted back to uranium and used as nuclear fuel.

Relative atomic mass

Most elements contain a mixture of isotopes. This explains why their relative atomic masses are *not* whole numbers.

> *The relative atomic mass of an element is the average mass of one atom, taking account of its isotopes and their relative proportions.*

For example, the mass spectrometer trace in figure 2 shows that chlorine consists of two isotopes with mass numbers of 35 and 37. These isotopes can be written as $^{35}_{17}Cl$ and $^{37}_{17}Cl$.

If chlorine contained 100% $^{35}_{17}Cl$, then its relative atomic mass would be 35. If it contained 100% $^{37}_{17}Cl$, then its relative atomic mass would be 37. A 50:50 mixture of $^{35}_{17}Cl$ and $^{37}_{17}Cl$ would have a relative atomic mass of 36. Figure 2 shows that naturally-occurring chlorine contains three times as much $^{35}_{17}Cl$ as $^{37}_{17}Cl$, i.e. 75% to 25%. This gives a relative atomic mass of 35.5, as shown in table 2.

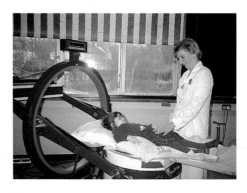

Radioactive isotopes are used in scanning equipment in hospitals

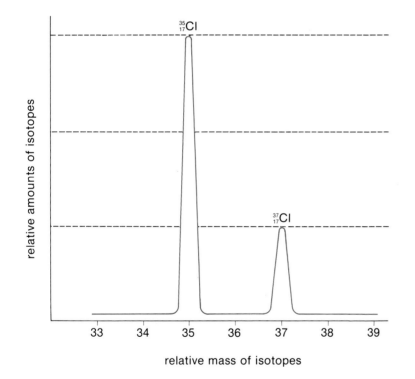

Figure 2
A mass spectrometer trace for chlorine. What are the relative amounts of $^{35}_{17}Cl$ and $^{37}_{17}Cl$?

Percentage of $^{35}_{17}Cl$	100	75	50	25	0
Percentage of $^{37}_{17}Cl$	0	25	50	75	100
Relative atomic mass	35	35.5	36	36.5	37

Table 2: The relative atomic mass of different mixtures of the isotopes of chlorine

Questions

1 There are three isotopes of hydrogen with mass numbers of one, two and three. (Naturally-occurring hydrogen is almost 100% $^{1}_{1}H$.) How many protons, neutrons and electrons do each of the three hydrogen isotopes have?

2 Neon has two isotopes, with mass numbers of 20 and 22.

(a) How would you expect the boiling point of $^{20}_{10}Ne$ to compare with that of $^{22}_{10}Ne$? Explain your answer.

(b) Suppose a sample of neon contains equal numbers of the two isotopes. What is the relative atomic mass of neon in this sample?

(c) Neon in the air contains 90% of $^{20}_{10}Ne$ and 10% of $^{22}_{10}Ne$. What is the relative atomic mass of neon in the air?

3 Discuss the following questions with two or three others.

(a) Why do isotopes have the same chemical properties, but different physical properties?

(b) Why do samples of natural uranium from different parts of the world have slightly different relative atomic masses?

(c) Why can chlorine form molecules with three different relative molecular masses?

5 Electron Structures

Chemical reactions involve changes in the number of electrons belonging to atoms. This led chemists to suggest that the noble gases must have very stable electron structures because they are so unreactive. This means that atoms or ions will have stable electron structures if they have 2 electrons (like helium), 10 electrons (like neon), 18 electrons (like argon), etc.

Chemists think that electrons occupy the outer parts of atoms in layers or **shells**. The first shell is filled and stable when it contains two electrons like helium. The second shell is filled when it contains eight electrons. So neon with 10 electrons has 2 electrons in the first shell and eight electrons in the second shell. We say that its electron structure is 2, 8. The third shell is also stable when it contains 8 electrons. So, argon is stable because its first, second and third shells are all filled with 2, 8 and 8 electrons respectively. We write the electron structure of argon as 2, 8, 8.

Figure 1 shows the first 20 elements arranged as in the periodic table. The electron structure of each element is written below its symbol. When the first shell is full at helium, electrons go into the second shell. So the electron structure of lithium is 2, 1; beryllium is 2, 2; boron is 2, 3; etc. When the second shell is full at neon, electrons start to fill the third shell and so on. Using these electron structures, we can explain why elements in the same group have similar properties.

Figure 1 (right)
The electron structures of the first 20 elements in the periodic table

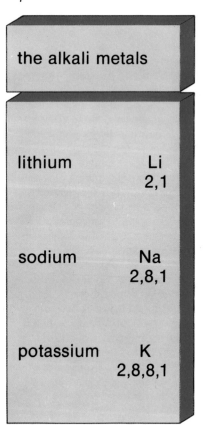

Figure 2
Electron structures of the alkali metals

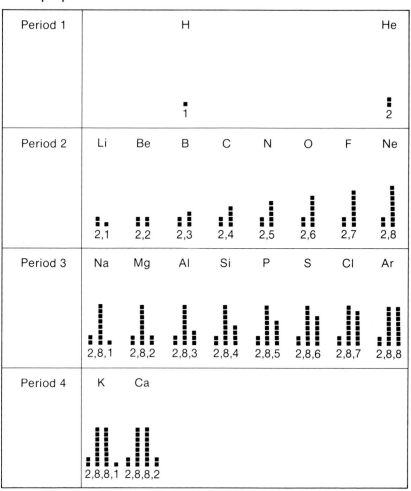

Group I: the alkali metals

Elements in the same group have similar electron structures. Each alkali metal follows a noble gas in the periodic table. So alkali metals have one electron in their outer shell (figure 2). By losing this one electron, their atoms form positive ions (Li^+, Na^+, K^+) with a stable electron structure like a noble gas. For example, the electron structure of Na^+ is 2, 8 which is like neon. So alkali metals have similar properties because they have similar electron structures.

- They are very reactive, being keen to lose the single electron in the outer shell.
- They all form ions with a charge of 1+; so the formulas of their compounds are similar.

Group VII: the halogens

Halogens come just before a noble gas in the periodic table. So, halogen atoms have seven electrons in their outer shells (figure 3). By gaining one electron, they form negative ions (F^-, Cl^-, Br^-) with stable electron structures like the next noble gas.

- They are reactive being so keen to gain one electron.
- They form ions with a charge of 1−, so the formulas of their compounds are similar.

Look at the table. It shows the electron structures of atoms and ions for elements in period 3.

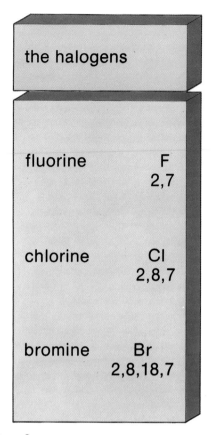

Figure 3
Electron structures of the halogens

Element	Na	Mg	Al	Si	P	S	Cl	Ar
Electron structure	2,8,1	2,8,2	2,8,3	2,8,4	2,8,5	2,8,6	2,8,7	2,8,8
Electrons in outer shell	1	2	3	4	5	6	7	8
Common ion	Na^+	Mg^{2+}	Al^{3+}	—	—	S^{2-}	Cl^-	—
Electron structure of ion	2,8	2,8	2,8	—	—	2,8,8	2,8,8	—

Electron structures of the atoms and ions of elements in period 3

1 The first three elements (sodium, magnesium and aluminium) *lose* the electrons in their outer shell to form positive ions (Na^+, Mg^{2+}, Al^{3+}) with an electron structure like the previous noble gas.

2 Sulphur and chlorine near the end of the period *gain* electrons to form negative ions (S^{2-} and Cl^-) with an electron structure like the next noble gas, argon.

3 Elements in the middle of the period, like silicon and phosphorus, do not usually form ions. They get stable electron structures in their compounds by *sharing* electrons with other atoms instead of gaining them or losing them. This sharing of electrons results in covalent bonds between atoms. It is the usual type of bonding in non-metal compounds. *So, during most reactions, atoms either lose, gain or share electrons in order to get a stable electron like a noble gas.* This idea forms the basis of the electronic theory of chemical bonding.

Questions

1 Why do all the alkali metals have similar chemical properties?

2 (a) Write down the electron structures of magnesium and calcium.
(b) How many electrons are there in the outer shell of an atom of an element in Group II?
(c) What charge will stable ions of Group II elements have?
(d) How many electrons are there in the outer shell of oxygen atoms?
(e) What is the charge on stable ions of oxygen?
(f) Why does magnesium react with oxygen to form an ionic compound, $Mg^{2+}O^{2-}$?

3 How many protons, neutrons and electrons will the following have? (i) N; (ii) N^{3-}; (iii) Ca; (iv) Ca^{2+}.

6 Ionic and Covalent Bonds

Ionic bonds: transfer of electrons

Figure 1 shows what happens when sodium chloride (Na^+Cl^-) is formed from sodium and chlorine atoms.

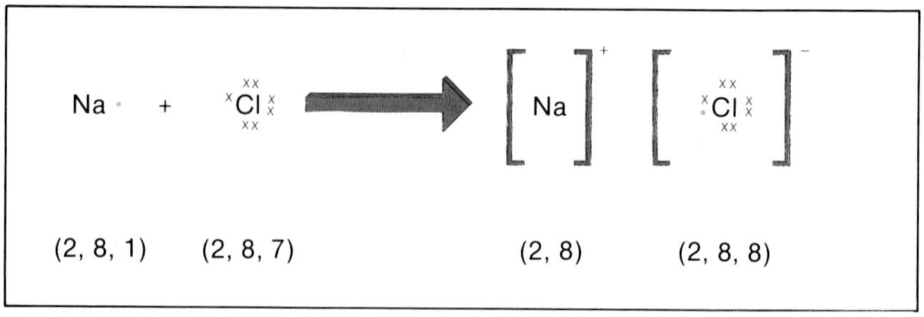

Figure 1
Electron transfer during the formation of sodium chloride

$$Na\cdot \quad + \quad \overset{\text{x x}}{\underset{\text{x x}}{^{\text{x}}Cl^{\text{x}}}} \quad \longrightarrow \quad \left[Na \right]^+ \quad \left[\overset{\text{x x}}{\underset{\text{x x}}{^{\text{x}}Cl^{\text{x}}}} \right]^-$$

(2, 8, 1)   (2, 8, 7)   (2, 8)   (2, 8, 8)

The number of electrons in the outer shell of each atom is shown by dots or crosses round its symbol. Full electron structures are also shown below the symbols. Each sodium atom loses the one electron in its outer shell to form a Na^+ ion with the same electron structure as neon. The electrons given up by sodium atoms are taken by chlorine atoms. Each chlorine atom gains one electron to form a Cl^- ion with the same electron structure as argon. So, the formation of NaCl involves the *complete transfer* of an electron from a sodium atom to a chlorine atom, forming Na^+ and Cl^- ions.

Ionic (electrovalent) bonds result from the attraction between these oppositely charged ions.

Compounds containing ionic bonds are called **ionic compounds**. The structure, bonding and properties of ionic compounds are discussed in units H8 and H9.

Figure 2 shows the electron transfer that takes place in the formation of magnesium sulphide and lithium oxide. Transfer of electrons to form ionic bonds is typical of the reactions between metals and non-metals.

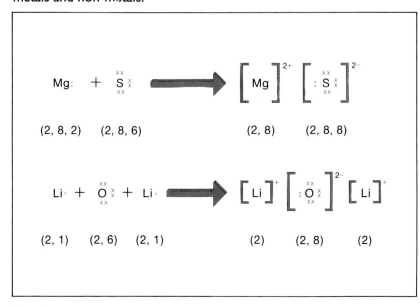

Figure 2
Electron transfers in the formation of magnesium sulphide and lithium oxide

Covalent bonds: sharing electrons

A chlorine atom is very unstable. Its outer shell contains only seven electrons. At normal temperatures, chlorine atoms join up in pairs to form Cl_2 molecules. Why is this? If two chlorine atoms come close together, the electrons in their outer shells can overlap so that one pair of electrons is shared by each atom (figure 3). The shared pair of electrons is attracted by the positive nucleus of each atom forming a **covalent bond**. The shared pair contributes to the outer shell of both the chlorine atoms. Circles are used to enclose the electrons in the outer shell of each chlorine atom. So,

> *A covalent bond is formed by the sharing of a pair of electrons between two atoms. Each atom contributes one electron to the bond.*

The atoms in molecular compounds are joined by covalent bonds. The structure, bonding and properties of these compounds are discussed in units H5, H6 and H7. The electron structures of some common molecular substances are shown in figure 4. Notice the following points.

1 All the atoms have an electron structure like a noble gas.

2 The electron structures of these molecular compounds can be related to their structural formulas, which show bonds as lines between atoms (e.g. H—O—H for water). The number of lines to an atom equals its valency and each line represents a shared pair of electrons.

3 Double covalent bonds result from the sharing of two pairs of electrons as in oxygen and carbon dioxide. Triple covalent bonds with three pairs of electrons are also known.

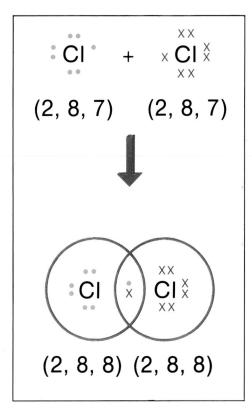

Figure 3
Electron sharing in the covalent bond in a chlorine molecule

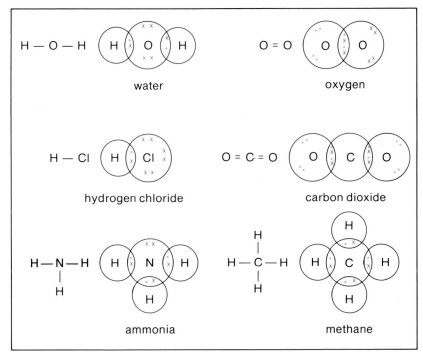

Figure 4
The electron structures of some simple molecular substances

Questions

1 Look at figure 2.
(a) How many electrons do magnesium and sulphur atoms lose or gain in forming magnesium sulphide, MgS?
(b) Why are these numbers of electrons lost or gained?
(c) How many electrons do lithium and oxygen atoms lose or gain in forming lithium oxide, Li_2O?
(d) Which noble gases have electron structures like the ions in lithium oxide?
(e) Why do two lithium atoms react with one oxygen atom in forming lithium oxide?

2 Element X has the electron structure 2, 6 and element Y has the electron structure 2, 8, 3.
(a) Write an equation similar to those in figure 2 for the formation of a compound between X and Y.
(b) What is the formula of this compound?

7 Radioactivity

Henri Becquerel—the discoverer of radioactivity

1896: Becquerel discovers radioactivity

In 1896, the Frenchman Henri Becquerel was investigating the reactions of uranium salts. Most of his experiments were carried out in bright sunlight. By chance, Becquerel left some uranium salt *in the dark* and on a photographic plate wrapped in black paper. When he developed the photographic plate, Becquerel was very surprised to see that it had been darkened (fogged) in the area near the uranium salt. Further experiments showed that the uranium salt was giving off some kind of radiation. Becquerel called the process **radioactivity** and described the uranium salts as **radioactive**. He also suggested that the radiation was a form of energy like light. This would explain why it caused a reaction on the photographic plate.

1898: Marie Curie discovers radium

Two of Becquerel's colleagues, Marie and Pierre Curie, decided to examine the radioactivity of uranium salts in more detail. They found that all uranium salts showed radioactivity and affected photographic plates. They also found that radioactive substances would either increase or decrease the charge on an electroscope. This meant that the radiation must carry a charge. During their investigations the Curies noticed that pitchblende (impure uranium sulphide) was much more radioactive than they had expected from its uranium content. This led them to think that the pitchblende contained another element more radioactive than uranium.

Marie Curie (1867–1934) and her husband Pierre (1859–1906). The Curies spent nearly four years isolating the radioactive elements radium and polonium from pitchblende. In 1903 they shared the Nobel Physics Prize with Becquerel. Then in 1911, Marie was awarded the Chemistry Prize—the first person to win two Nobel Prizes. In 1934, Marie died of cancer, probably caused by the radioactive materials she had spent her life studying

In 1898, after months of hard work, Marie Curie isolated two other radioactive elements from pitchblende. She named these elements radium (after the term 'radioactivity') and polonium (after Poland where she was born). Radium was found to be two million times more radioactive than uranium.

The fogging of a photographic plate by radioactive materials has practical applications today. For example, we can follow the way in which a plant leaf takes in carbon dioxide by exposing the leaf to carbon dioxide containing radioactive carbon-14. After some time, the leaf is held against a photographic plate. When the plate is developed, the darkest parts show where the carbon-14 has gone.

Radioactivity can be used to study the uptake and distribution of metal ions in plants. In this photo the plant has been grown in a solution of a radioactive mercury compound for 48 hours. The radioactive mercury can then be detected in the stem and leaves by laying the plant on photographic film and allowing sufficient exposure time

Before very long, another effect of radioactive substances was discovered. When a screen coated with zinc sulphide is placed near a radium salt and examined with a magnifying lens, tiny flashes of light appear on the surface of the zinc sulphide. The flashes of light are called **scintillations**. When particles hit the zinc sulphide, their kinetic energy is converted to light which causes the scintillations. The particles are produced as the radioactive radium atoms break up.

The radiations and particles emitted by radioactive substances were originally detected in one of three ways:

1 the fogging of photographic plates;
2 the discharge of an electroscope;
3 scintillation methods.

Nowadays, the radiations are detected using a Geiger-Müller tube (GM tube) connected to a counter or ratemeter. This is discussed in unit 9 of this section.

Questions

1 (a) Why are photographic plates darkened ('fogged') when they are left in the dark near uranium salts?
(b) Why did the Curies think that pitchblende contained an element more radioactive than uranium?
(c) Why do some radioactive materials cause the discharge of an electroscope?
2 (a) What are scintillations?
(b) What causes the scintillations when a radium salt is placed near zinc sulphide?
3 (a) Find out about the personal life of Marie Curie.
(b) Find out about Marie Curie's scientific work.
(c) Why do you think that Marie Curie is one of the most admired scientists of the 20th century?

8 Nuclear Reactions

Alchemists dreamed of changing base metals to gold

Radioactive radium compounds are used in the luminous paints on the hands and numbers of clocks and watches

Mixtures of zinc sulphide and a small amount of any radium salt glow in the dark. These mixtures are used in luminous paints on the hands and numbers of clocks and watches. We now know that the glow is caused by **alpha particles** hitting the zinc sulphide in the paint. This stops the alpha particles and their kinetic energy is converted to light energy. The alpha particles are produced as radium atoms split up (disintegrate) of their own accord. All the time, radium atoms are breaking up and losing alpha particles. This spontaneous break up of atoms results in **radioactivity (radioactive decay)**. Radioactivity arises from the breakdown of unstable nuclei. Alpha particles are helium ions, $_2^4\text{He}^{2+}$—helium atoms that have lost both of their electrons. The mass number of an alpha particle is four and its atomic number is two. Thus, when an atom of $_{88}^{226}\text{Ra}$ loses an alpha particle, the fragment left behind will have a mass number which is four less than $_{88}^{226}\text{Ra}$ and an atomic number which is two less than $_{88}^{226}\text{Ra}$. So, the fragment will have a mass number of 222 and an atomic number of 86. All atoms of atomic number 86 are those of radon, Rn. Thus, the final products are $_{86}^{222}\text{Rn}$ and $_2^4\text{He}$. The nuclear decay for radium-226 can be summarized in a nuclear equation as

$$_{88}^{226}\text{Ra} \rightarrow {}_{66}^{222}\text{Rn} + {}_2^4\text{He}$$

Radioactive decay with the loss of an alpha particle is common for large isotopes with an atomic number over 83. These include uranium-238, radium-226 and plutonium-238. These isotopes decay because they are simply too heavy. They lose mass and try to become stable by losing an alpha particle.

Radioactive isotopes with atomic numbers below 83 usually lose **beta particles** when they decay, rather than alpha particles. Experiments show that beta particles are electrons, $_{-1}^{0}e$. Carbon-14 ($_{6}^{14}C$) decays by losing beta particles to form $_{7}^{14}N$. The decay process can be represented as

$$\text{Mass} \searrow \quad \text{Charge} \nearrow \quad _{6}^{14}C \rightarrow {_{7}^{14}N} + {_{-1}^{0}e}$$

Notice that the *total* mass and the *total* charge are the same after beta decay as they were before. During beta decay, a neutron in the nucleus of the radioactive atom splits up into a proton and an electron. The proton stays in the nucleus, but the electron is ejected as a beta particle. Thus, the mass number of the remaining fragment stays the same, but its atomic number increases by one (see the table).

	Isotopes involved	Particle lost	Change in mass number	Change in atomic number
Alpha decay	Atomic number usually >83	$_{2}^{4}He^{2+}$	−4	−2
Beta decay	Atomic number usually ⩽83	$_{-1}^{0}e$	0	+1

Comparing alpha decay and beta decay

Chemical reactions and nuclear reactions

Chemical reactions involve changes in the electrons in the outer parts of atoms. During chemical reactions electrons are either

1 *transferred* from one atom to another; or
2 *shared* between two atoms.

Nuclear reactions, on the other hand, involve changes in the nuclei in the central parts of atoms. During nuclear reactions, one element may be converted to another element by

1 *radioactive decay*;
2 *atomic fission* (unit 11); *or*
3 *atomic fusion*.

The first scientist to show that one element could be converted to another was Rutherford. In the 1920s, Rutherford and his colleagues turned sodium into magnesium and aluminium into silicon. At first, the changes from one element to another involved only elements with low mass numbers. However, in 1940, American chemists managed to build up heavier elements from uranium, the heaviest natural element. For example, neptunium-239 and plutonium-239 were obtained by bombarding uranium-238 with neutrons. By 1972, thirteen new elements had been synthesized, all of which come after uranium in the periodic table. All these new elements are radioactive and most of them disintegrate rapidly.

9 Detecting Radioactivity

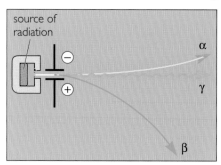

Figure 1
The effect of an electric field on different ionizing radiations

When radioactive substances decay, they emit alpha particles ($^4_2\text{He}^{2+}$ ions) and beta particles (electrons). At the same time, they usually give off **gamma rays**. Gamma rays are electromagnetic waves, like light. Their properties are very similar to those of X-rays. Gamma rays contain so much energy that they can damage cells and kill organisms. As a radioactive atom gives off gamma rays, it loses energy and becomes more stable.

When alpha particles, beta particles and gamma rays pass through air and other materials, they cause particles in the materials to lose electrons and form ions. Because of this, they are sometimes called **ionizing radiations**. Although ionizing radiations can be harmful, they do have beneficial uses. These uses are discussed in unit 10.

When alpha particles are emitted by radioactive substances, they travel a few centimetres in air and they can be deflected by electric and magnetic fields. Beta particles are much lighter than alpha particles. They travel several metres in air and are deflected much further by electric and magnetic fields. Gamma rays are unaffected by electric and magnetic fields, but they travel a long way in air and penetrate bricks and metal sheets. The properties of alpha particles, beta particles and gamma rays are summarized in table 1 and figures 1 and 2.

Table 1: The nature and properties of alpha particles, beta particles and gamma rays

Radiation	Nature	Effect of electric and magnetic fields	Distance travelled in air	Penetration of		
				paper	thin aluminium	thick lead
Alpha	helium nuclei, $^4_2\text{He}^{2+}$	small deflection	a few cms	✗	✗	✗
Beta	electrons, $^0_{-1}\text{e}$	large deflection	a few metres	✓	✗	✗
Gamma	electromagnetic waves	no deflection	many kms	✓	✓	✗

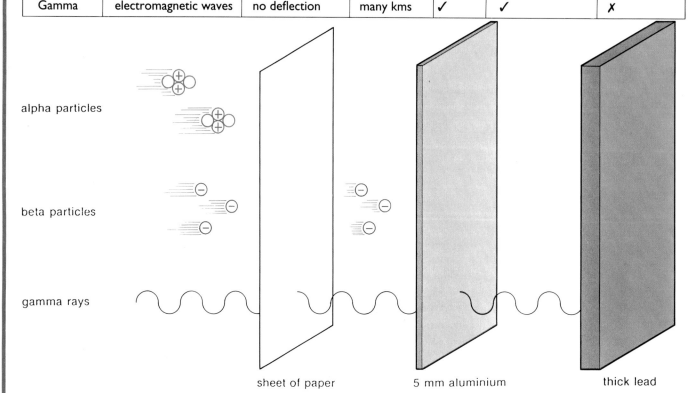

Figure 2 *The relative penetrating power of alpha particles, beta particles and gamma rays*

Background radiation

Various rocks in the Earth, including granite, contain small amounts of radioactive uranium, thorium and potassium compounds. Our bodies also contain traces of radioactive materials as do the bricks which are used to build our homes, schools and factories. In addition to these sources, we are also exposed to radiation from the Sun. These four *natural* sources of radiation are called **background radiation**. We are exposed to this background radiation all our lives. Normal background radiation is very low. It causes no health risk.

Half life

The rate of decay of a radioactive isotope is shown by its **half life**. *This is the time it takes for half of the atoms of the isotope to decay.* Half lives can vary from a few milliseconds to several million years. The shorter the half life, the faster the isotope decays and the more unstable it is. The longer the half life, the slower the decay process and the more stable the isotope. Uranium-238 with a half life of 4500 million years is 'almost stable', but polonium-234, with a half life of only 0.15 milliseconds is quite the reverse.

The radioactive isotope iodine-131 has been used by doctors to measure the uptake of iodine by the thyroid, an important gland in the human body. Iodine-131 has a half life of eight days. This means that, starting with 1 g of iodine-131, only ½ g remains eight days later. After another eight days, the half gram will have decayed to ¼ g and eight days after that (i.e. 24 days from the start) only ⅛ g would remain (figure 3). In fact, only tiny quantities of iodine-131 are used in this procedure. By measuring the amount of radioactivity in the thyroid gland, doctors can tell how much of the iodine-131 has been taken up by the gland.

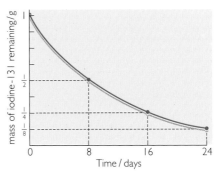

Figure 3
A radioactive decay curve for I-131. The graph shows the mass of I-131 remaining when 1 g of the isotope decays for 24 days

Questions

1 In the UK, an average of 78% of the total radiation received by a person comes from background radiation. This is composed of 16% from rocks and soil, 16% from natural radioactive substances in our bodies, 33% from bricks and other building materials and 13% radiation from the Sun.

(a) Use these figures to draw a pie chart showing where the total radiation comes from.

(b) What percentage of the total radiation to which we are exposed comes from artificial sources?

(c) What do you think are the main sources of this artificial radiation?

2 The figure shows a gamma ray unit used in a hospital. **A** is a container of radioactive material. The unit is made 'safe' by moving the container to

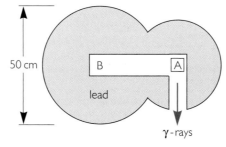

position **B**. The radioactive material is cobalt-60 ($^{60}_{27}$Co). Cobalt-60 has a half life of 5 years.

(a) Why is the unit made mainly of lead?

(b) What is (i) the atomic number, (ii) the mass number of $^{60}_{27}$Co?

(c) How many protons, neutrons and electrons are there in one cobalt-60 atom?

(d) Why is the gamma ray unit 'safe' when the container is in position **B**?

(e) Suggest *three* safety precautions which someone should take when removing a container of cobalt-60 from the unit?

3 (a) What do you understand by the term 'radioactive element'?

(b) Name two radioactive elements which occur naturally.

(c) Which part of an atom is responsible for radioactivity?

(d) Which of the particles or rays emitted by radioactive substances (i) is most penetrating, (ii) contains positive particles, (iii) is not deflected by a magnetic field?

10 Using Radioactive Isotopes

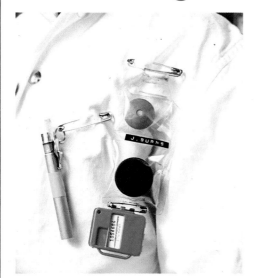

A technician wearing a blue radiation-level badge

Radioactive isotopes are widely used in industry and medicine. People who work with radioactive materials must wear a radiation badge which measures the radiation to which they have been exposed. Gamma rays are the most penetrating and dangerous form of radiation. They can cause changes in the structure of chemicals in our bodies and kill living cells. Scientists and technicians who use dangerous isotopes must be protected by lead or concrete shields and handle radioactive materials by remote control.

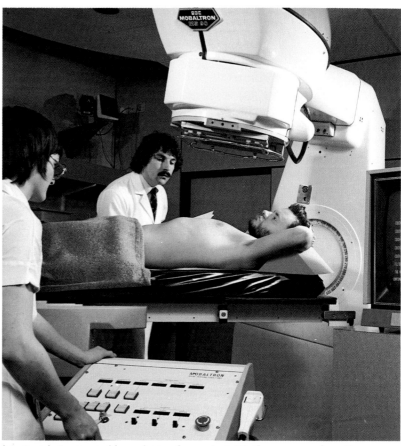

A patient being treated by radiation from cobalt-60. Gamma rays from the cobalt-60 can penetrate the body and kill cancer cells

Medical uses

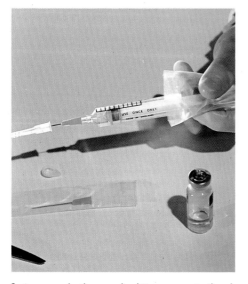

Syringes and other medical items are sterilized by gamma radiation. In some countries, gamma radiation is used to prevent food going bad

When a living cell is exposed to radiation, the structure of chemicals (genes) in the nucleus may be changed. This can cause the cell to die. Penetrating gamma rays from cobalt-60 ($^{60}_{27}$Co) are used to kill cancer cells and to treat growths inside the body. Cancer cells on the surface of the body, as in skin cancer, can be treated with less penetrating beta rays. This is done by strapping a plastic sheet containing $^{32}_{15}$P (radioactive phosphorus) or $^{90}_{38}$Sr (radioactive strontium) on the affected area. Cancer can be fatal, but doctors are becoming more and more successful at curing less serious cases.

Medical items, such as dressings and syringes, are sealed in polythene bags and sterilized by gamma rays. This method of sterilization is much easier than the old method which used steam. Intense doses of gamma radiation sterilize the articles by killing any bacteria on them.

Tracer studies

Radioactive isotopes are easy to detect. Therefore they are used to *trace* what happens to different substances in chemical, physical and biological processes. Radioactive isotopes are usually mixed with non-radioactive atoms of the substance under investigation. For example, the way that plants take in and use phosphates can be studied using a fertilizer containing $^{32}_{15}P$. Tracer studies using $^{14}_{6}C$ have helped in the study of photosynthesis and protein synthesis. The activity of the thyroid gland can be studied by measuring the uptake of iodine-131 (unit 9 of this section).

Archaeological and geological uses

The common isotope of carbon is carbon-12. The carbon in living things is therefore mainly carbon-12 with a small constant percentage of *radioactive* carbon-14. This gets into living things from the carbon-14 which is part of the carbon dioxide in the air. When an animal or plant dies, the carbon-14 in it continues to decay. However, the replacement of decayed carbon-14 from food and carbon dioxide stops. The amount of carbon-14 left in the remains of the animal or plant can be measured. Then, knowing the half life of carbon-14, it is possible to work out how long it is since the animal or plant died. 'Carbon dating' has been used to check the age of ancient documents and the bones of ape men. Using a technique similar to carbon dating, geologists can work out the age of rocks. Most rocks contain at least traces of radioactive U-238. This decays over millions of years in a series of reactions to form lead-206. If we assume that the uranium bearing rock contained no lead-206 originally, then the present ratio of U-238 to Pb-206 in the rocks can be used to calculate the time since the rocks formed.

Part of the Dead Sea Scrolls. Radiocarbon dating shows that the Dead Sea Scrolls are about 2000 years old and probably authentic

---

Questions

1 What precautions are taken to ensure that scientists and technicians who work with radioactive isotopes are not exposed to dangerous radiations?

2 Suppose that carbon-14 makes up *x*% of the carbon in living things and that the half life of carbon-14 is 5700 years.
  (a) How long will it take for the percentage of carbon-14 to fall from:
  (i) *x*% to ½*x*%, (ii) ½*x*% to ¼*x*%?
  (b) How old is an object which has ⅛*x*% of its carbon as carbon-14?

3 Why are radioactive isotopes important?

4 In some countries, gamma radiation is used to prevent food going bad.
  (a) What will gamma radiation do to bacteria to prevent the food going bad?
  (b) This process is not yet used in the UK. Why do you think that some people are opposed to its use?

5 Suppose the half-life for the decay of U-238 to Pb-206 is 4×10^9 years. What is the age of a rock in which the ratio of U-238 to Pb-206 is (i) 1:1, (ii) 1:3?

6 A medical gamma ray unit was supplied with 10 g of caesium-137 which has a half life of 35 years.
  (a) How much caesium-137 would be left after (i) 35 years, (ii) 70 years, (iii) 140 years?
  (b) When will its rate of decay be half as great as it was at the start? Explain your answer.
  (c) Suppose the medical unit is closed after 20 years. How much caesium-137 is left at this time? (Hint: it may help you to draw a graph.)

11 Nuclear Energy

Enormous amounts of energy can be obtained from nuclear reactions, but this discovery was made almost by chance. In 1938, the German scientists Hahn and Strassmann were trying to make a new element by bombarding uranium with neutrons. Instead of producing a new element, the neutrons caused the uranium nuclei to break up violently into two smaller nuclei. Two or three separate neutrons were also released during each reaction together with large quantities of energy (figure 1).

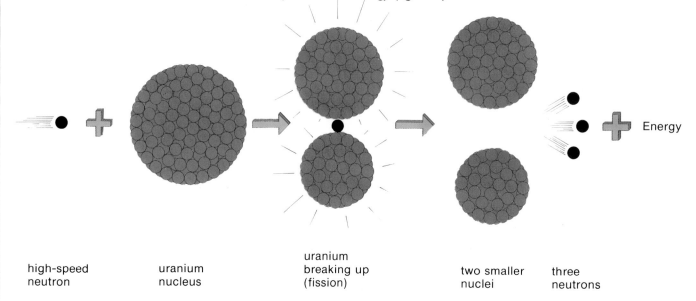

| high-speed neutron | uranium nucleus | uranium breaking up (fission) | two smaller nuclei | three neutrons |

Figure 1
The fission of a nucleus of uranium-235

Notice how this nuclear reaction differs from radioactive decay. First, it does not happen of its own accord like radioactive decay. It only happens when the uranium is bombarded by neutrons. Second, it involves the break up of one large nucleus into two fragments of roughly the same size. During radioactive decay, the products are one large fragment and one very small fragment (either an alpha particle or an electron). The special name **nuclear fission** (atomic fission) is used to describe the splitting of an atom into two fragments of roughly the same size.

Atomic bombs and atomic reactors

Natural uranium contains two isotopes, $^{235}_{92}U$ and $^{238}_{92}U$. Only 0.7% is uranium-235. Experiments showed that only uranium-235 took part in nuclear fission as a result of neutron bombardment. This led scientists to realize that if the uranium sample contained a larger fraction of uranium-235, the neutrons released during fission would split more uranium nuclei and cause a chain reaction.

Figure 2 shows how an exploding chain reaction occurs in uranium-235. One reaction releases three neutrons, three neutrons release nine, then 27, 81 and so on. Each time more and more energy is produced as more and more uranium-235 atoms undergo fission.

In an atomic bomb, the fission of *enriched* uranium-235 happens in an *uncontrolled* manner and enormous amounts of energy and radiation are released. In an atomic reactor, a *mixture* of uranium-235 and uranium-238 is used to give a *controlled* chain reaction. In this

Atomic bomb explosions release vast amounts of energy

case, the heat produced is used to generate electricity. Figure 3 shows a simplified diagram of a gas-cooled nuclear reactor. At the centre of the reactor, rods of 'enriched' uranium containing 3% of uranium-235 are stacked inside a large block of graphite. Neutrons released by the fission of the uranium are slowed down by collision with carbon atoms in the graphite.

The temperature of the reactor is controlled by moveable rods of boron or cadmium which absorb neutrons. The deeper these rods are inserted, the more neutrons are absorbed and the slower the reaction becomes. By carefully adjusting the neutron-absorbing rods, a controlled chain reaction can be obtained. Heat is produced steadily and taken away by carbon dioxide gas circulating through the reactor. The hot gas is then used to make steam which drives turbines and generates electricity. In the USA, most reactors use water as a coolant in place of carbon dioxide.

Recently, reactors have been developed that use plutonium as the fuel and liquid sodium as the coolant. Unlike uranium, which reacts best with slow neutrons, plutonium can use *fast* neutrons. This means that the reactor does not need a graphite block to slow them down. These reactors are called fast reactors. The first power station to use a fast reactor for generating electricity was the one at Dounreay in Scotland.

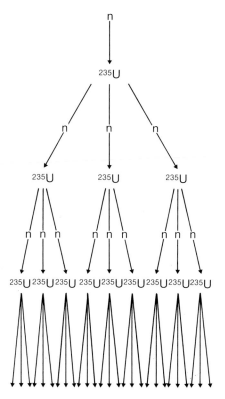

Figure 2
A chain reaction in uranium-235

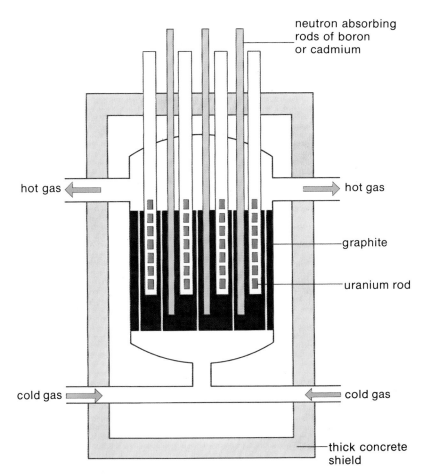

Figure 3
A simplified diagram of a gas-cooled nuclear reactor

Questions

1 Explain the following terms in your own words.
nuclear fission; *nuclear energy*; *chain reaction*; *fast reactor*?
2 What are the differences between nuclear fission and radioactive decay?
3 In what ways are nuclear reactors similar to nuclear bombs? In what ways are they different?
4 This question is about nuclear reactors?
  (a) Why are the uranium rods sunk in a graphite block?
  (b) What is the function of the boron or cadmium rods?
  (c) How is the rate of nuclear fission controlled?
  (d) Why is the reactor surrounded by thick concrete?
  (e) How is the energy from nuclear fission converted into electricity?
  (f) Recently, some reactors have used liquid sodium as the coolant. Mention one advantage and one disadvantage of using sodium.

Section K: Activities

The nuclear debate

Hiroshima, in Japan, a few weeks after the atomic bomb exploded in August 1945. The long wooden building in the foreground was, of course, erected after the blast

Hardly a week passes without some mention of nuclear technology in the newspapers or on television. A lot of the publicity concerns the dangers involved. These include pollution by radioactive waste and, of course, the horrors of nuclear war.

The first atomic reactor was built in America in 1942 by the Italian scientist, Enrico Fermi. The first industrial nuclear power station was opened in 1956 at Calder Hall, now Sellafield, in Cumbria, UK.

The first nuclear (atomic) bomb was tested in July 1945. Robert Oppenheimer, who had played a large part in the bomb's development, was shocked by its enormous power. Afterwards, he said, 'we knew the world would not be the same'. A month later a nuclear bomb was dropped on Hiroshima in Japan, killing over 70 000 people.

Oppenheimer was opposed to the development of the hydrogen bomb, which was even more destructive than the nuclear bomb. Nevertheless, a hydrogen bomb was tested in the Pacific Ocean in 1953. Many people were worried about the possibility of such weapons being used in a war and large numbers of people joined CND (the Campaign for Nuclear Disarmament).

As more nuclear reactors are built, there is an increasing danger of pollution by radioactive waste. After use, the spent fuel is still very radioactive. Some of the waste material will remain dangerous for hundreds of years. Several methods have been used to dispose of radioactive waste. These include dumping the waste in deep mines or solidifying it in glass or concrete and then burying it below the sea bed.

Many people are also concerned about the link between nuclear power stations and nuclear weapons. They fear that terrorists might steal uranium or plutonium and use it to make bombs. There is also the risk that terrorists might capture a nuclear power station and use

CND marchers carrying banners and CND slogans. During the 1960s, 1970s and 1980s, the USA and the USSR continued to develop more and more powerful nuclear weapons. This led to more support for CND and the anti-nuclear movement. During the 1990s, the 'thawing of the cold war' and improvements in East–West relations has eased the tensions and resulted in less CND activity

it as a bargaining device by threatening to destroy the reactor.

In 1986, a nuclear reactor exploded at Chernobyl in Russia. 31 people died in the explosion and 135 000 people were evacuated from their homes. Clouds of radioactive materials were carried by winds to all parts of Europe. The contamination led to restrictions on the sale of meat and vegetables as far away as Scotland and Wales.

The main causes of the accident at Chernobyl were serious faults in the reactor design. This type of reactor is used only in the Soviet Republics. As a result of the accident, the design and the safety measures in these reactors has been improved. A reactor like the one at Chernobyl could not be built in the UK because it does not meet our safety requirements.

Many people feel that the problems created by nuclear technology outweigh the advantages. Other people say that we must learn to live with the risks if we are to maintain our industrial activity, preserve our standards of welfare and improve the quality of life for those in the Third World.

Any discussions on nuclear technology are always linked to energy supplies. The world consumption of coal, oil and gas will continue to increase. This is particularly so in developing countries. In the long term, we must find another large source of energy and the only possibility seems to be nuclear power. Solar energy, tidal energy and wind energy may help in certain areas, but none of them could provide the huge amounts of energy that we need.

Unfortunately, the reserves of uranium are also limited. Some experts think that there is not enough for the next century. Eventually, we may be forced to build more fast reactors that use plutonium rather than uranium. These can extract 50 to 60 times more energy from the fuel than ordinary uranium reactors.

Some radioactive waste used to be sealed in drums and then dumped at sea. There has been a world-wide agreement to stop this. Instead, waste is to be buried deep under ground on land

When the nuclear reactor exploded at Chernobyl in Russia, radioactive material was carried on the wind all over Europe. Many people say accidents like this make nuclear power too risky, but others believe we will not be able to supply the world's future energy needs without it

Questions

1 (a) What methods are used to dispose of waste from nuclear power stations?
   (b) Why are such thorough methods of disposal needed?
2 Why do some people think that we will have to rely on nuclear energy in the future?
3 Design a poster to show the main dangers associated with nuclear technology.

4 Is nuclear power the answer to our energy problems or are the risks too great? What do you think?
Plan a questionnaire and a survey to find out what other people think.
   (a) Decide the questions for your survey.
   (b) Carry out your survey.
   (c) Prepare a report of your survey.

1 Question 1 concerns the five particles labelled *A* to *E* below.

    A alpha particle
    B beta particle
    C hydrogen atom
    D proton
    E neutron

Which of these particles
(a) would be attracted by a positive plate?
(b) is the heaviest?
(c) causes fission of uranium-235?
(d) could be used in a thickness gauge for polythene or paper?
(e) contains the same number of protons and electrons?

2 Dalton put forward his Atomic Theory in 1807. The main points in his theory are listed below.

- All matter is made up of tiny particles called atoms.
- Atoms cannot be made or broken apart.
- All the atoms of one element are exactly alike.
- Atoms of one element are different from those of another element.
- Atoms combine in small numbers to form molecules.

Study each of these five points in turn.
(a) Say whether each point is still true or whether modern knowledge about atoms, molecules and atomic structure have shown them to be inaccurate.
(b) If a point is now inaccurate, explain why.

3 When a stream of alpha particles is directed at thin aluminium foil, most of the alpha particles pass straight through. The remaining alpha particles are either deflected or seem to bounce back off the foil. If the energy (speed) of the alpha particles is increased, some of them are absorbed by the aluminium.
(a) Why do you think most alpha particles pass straight through the foil?
(b) Why do you think some alpha particles are deflected as they pass through the foil?
(c) Why do you think some alpha particles seem to bounce back off the foil?
(d) What do you think happens to alpha particles which are absorbed by the aluminium?
(e) Why do you think the alpha particles must have high energy before they are absorbed by the foil?
(f) Write a nuclear equation for the process which occurs when high-energy alpha particles are absorbed by aluminium.

4 Radioactivity can be used in hospitals to investigate inside a patient's body without having to cut the person open. Technetium-99 (^{99}Tc) is one of the radioactive isotopes

which is used. This gives out gamma rays and has a half life of six hours.
(a) Why is lead used to shield hospital workers while they are using ^{99}Tc?
(b) Why is an isotope with such a short half life used?
(c) Use a periodic table to suggest what might happen to the ^{99}Tc when it decays.
(d) Medical equipment such as disposable syringes are sealed in polythene bags and then sterilized using gamma rays. Why is this a better method of sterilization than heating them to high temperatures in steam and then packing them?

5 The ion X^{2-} contains 54 electrons.
(a) To which group of the periodic table does the element *X* belong?
(b) Write the formula of the compound formed between caesium and *X*. (Caesium has symbol Cs and is in Group 1 of the periodic table.)
(c) How many protons are present in the nucleus of an atom of *X*?

SEG

6 The following passage was written to support the building of more nuclear power stations.

'World energy requirements will rise rapidly in the next decade. By the year 2000, it is estimated that the average world citizen will consume twice as much energy as the present average world citizen in the USA. This energy can only be obtained by building 4000 nuclear fission reactors using the inexhaustible supplies of nuclear fuel.

We must stop using fossil fuels to produce energy because they are irreplaceable as raw materials for the chemical industry. Their use also damages the environment. Atmospheric pollution is caused by burning coal and oil. Oil drilling and transport causes marine pollution and coal mining damages the environment.'
(a) Criticize the arguments used in the passage.
(b) Suggest four alternative sources of energy not considered in the passage.
(c) Discuss three important arguments against increasing the number of nuclear power stations.

7 Natural boron contains 20% boron-10 and 80% boron-11. Boron is the fifth element in the periodic table.
(a) What is the atomic number of boron?
(b) What are the mass numbers of the two boron isotopes?
(c) How many protons, neutrons and electrons does one atom of boron-11 possess?
(d) What is the relative atomic mass of boron?
(e) Samples of boron obtained from naturally-occurring boron compounds in different parts of the world have slightly different relative atomic masses. Why is this?

SECTION L
Earth and Atmosphere

1 Origins of the Earth and Atmosphere
2 Weathering and Erosion
3 Rocks in the Earth
4 Earth Movements
5 Plate Tectonics

The Earth's surface is continually being changed and weathered by temperature changes and the action of water

The Earth is a ball of rock and iron with a radius of about **6400 km.** About 4500 million years ago, the Earth was a mass of molten rock. Gradually, the Earth cooled down over millions of years. During this period, heavier metals sank to the centre of the Earth forming the present **core** of dense molten iron, sulphur and nickel at about 4000° C (figure 1). This core is surrounded by a thick layer of moderately dense solid and molten rock in the **mantle.** Temperatures in the mantle range from 1500 to 3500° C.

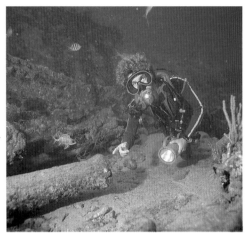

If we want to explore the highest or lowest points on the Earth's surface then we must use special breathing equipment

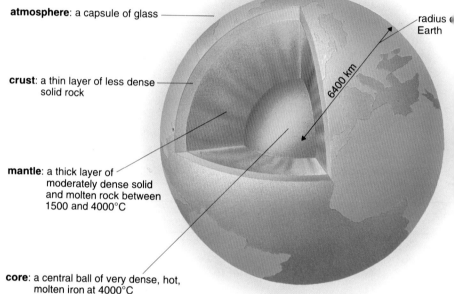

atmosphere: a capsule of glass

crust: a thin layer of less dense solid rock

radius of Earth

6400 km

mantle: a thick layer of moderately dense solid and molten rock between 1500 and 4000°C

core: a central ball of very dense, hot, molten iron at 4000°C

Figure 1
Layers of the Earth. Notice how the temperature and density of the layers increase towards the centre of the Earth

Less dense material collected on the surface of the Earth forming a thin **crust** about 50 km thick. Where the crust is thickest, its surface is above sea level.

Outside and above the Earth is the atmosphere—a layer of gases about 100 km deep. At about 6 km above sea-level, the air becomes too thin to survive. The oceans are about 6 km deep, so almost all life on Earth exists in a relatively thin band about 12 km thick.

Origins of the atmosphere and oceans

While the Earth was still forming, the gases above its surface were mainly hydrogen and helium. These gases had such small molecules that they escaped from the Earth's gravitational attraction into outer space. As the molten, volcanic surface cooled, other gases were added as rocks decomposed and elements reacted. These included water vapour, carbon dioxide, methane and nitrogen.

As the temperature dropped still further, water vapour condensed to form rivers, lakes and oceans.

Water vapour in the atmosphere condenses to form clouds as the air is forced to higher and colder zones by high mountains

When plants appeared on the Earth 3500 millions years ago, they used water and carbon dioxide for photosynthesis and added oxygen to the atmosphere. At the same time, plants used up the oxygen during respiration.

Flammable gases, like hydrogen and methane, burnt in this oxygen producing more water and carbon dioxide.

The composition of the atmosphere has remained more or less constant for the last 500 million years.

The main constituents are nitrogen and oxygen with smaller amounts of water vapour, carbon dioxide and noble gases (table 1). The Earth is the only planet in our solar system with oxygen in its atmosphere and abundant surface water in rivers, lakes and oceans. Other planets, such as Mars, do, however, have some water vapour and polar ice caps.

Nitrogen	78.09
Oxygen	20.95
Argon	0.93
Carbon dioxide	0.03
Noble gases	[Traces]
(Helium, Neon, Krypton and Xenon)	

Table 1: The percentages of gases in dry air

Maintaining the composition of the atmosphere

The composition of the atmosphere has remained constant for the last 500 million years due to processes in the carbon cycle (unit I4), the nitrogen cycle (unit J9) and the water cycle (unit B5).

In the carbon cycle,

• carbon dioxide and water vapour are removed from the atmosphere by photosynthesis but returned to the atmosphere when animals and plants respire and during the combustion of carbon compounds.

In the nitrogen cycle,

• nitrogen is removed from the air by nitrogen-fixing bacteria in plants and during the Haber Process, but returned to the air when animal and plant matter dies and decays.

In the water cycle,

• water vapour is continually being added to and removed from the atmosphere by processes of evaporation and condensation respectively. The solubility of gases in the water of rivers and oceans, even though it is very low for oxygen and nitrogen, also helps to maintain a balance of gases in the atmosphere.

Aerial photos such as this one taken above the Lake District help us to understand the formation of features on the Earth like rivers and deltas

Questions

1 (a) What substances do the Earth's surface and atmosphere contain which are essential for living things?
   (b) How did these substances get into the atmosphere and on the Earth's surface?
   (c) Why do you think there are no living things on:
   (i) the planet, Mercury,
   (ii) the planet, Pluto?
2 What activities of humans have begun to affect the composition of the atmosphere during the last 150 years?

2 Weathering and Erosion

The landscape is slowly and continually changing. Several processes act upon it, wearing down old rocks and creating new ones. These processes involve weathering, transport and deposition. They wear away mountains, cut out river valleys and deposit sediments in river deltas.

Weathering

The first stage in wearing down a landscape is weathering.

> *Weathering is the breaking down of materials by the atmosphere.*

The atmosphere attacks rocks and buildings in various ways so there are different kinds of weathering. Limestone reacts very slowly with rain water. Over hundreds of years the rock is worn away, creating limestone pavements, caves, sink holes and dry valleys (unit G10). Other rocks, like soft sandstone and chalk are weathered and worn away by the wind as well as the rain.

Hard rocks are not necessarily slower to weather than softer rocks. Granite, for example, is very hard, but one of its constituents (feldspar) reacts with water to form clay. As the feldspar reacts to form clay, the granite breaks up. Even the very hardest rocks, like quartz, are weathered in time. When the temperature falls below 0° C, ice forms. If the ice forms in cracks or crevices, it can break rocks apart because water expands as it freezes.

The expansion and contraction which take place as a result of temperature changes also results in the cracking, breaking and weathering of rocks.

This old statue and the new pillar on which it stands are both made of limestone. Notice the difference between the statue and its pillar. The old limestone statue has been weathered over the centuries by acid in the rain water

This photo shows a china clay pit in Cornwall. China clay (kaolin) is formed when feldspar in granite rocks weathers by reaction with water

Erosion

In mountainous areas, rocks and soil fall down steep slopes into valleys. Fast moving streams pick up some of this material and carry it away. This carrying away of weathered material is called **transport**.

The process of wearing away rocks and then carrying the weathered material away is called **erosion**.

i.e. Erosion = Weathering + Transport

The most important agents of erosion are moving water in rivers and waves, winds (especially in desert areas) and glaciers. These agents pick up particles of weathered material and transport them from one place to another.

A critical speed of flow of water, air (in winds) or ice (in glaciers) is needed before particles of a particular size can be picked up and transported. If the speed of flow falls below this critical level, then the particles will be deposited as sediment. This is called **deposition**. The heaviest material is deposited first and then the lightest.

The rock particles carried in the erosion agent wear away the landscape even further. The rock particles act like a scouring pad. The particles in rivers wear away the river bed, deepening and widening valleys. Coastal waves undercut cliffs and gouge out caves. Desert winds scrape rock surfaces and carry away weathered particles. Glaciers carve slowly and deeply as they move over the land.

The scree (broken rock) at the base of this rock face has formed as a result of temperature changes. Water collects in crevices in the rocks and expands when it becomes ice. Pieces of rock fall away when the weather becomes warmer and the ice melts

The soft sandstone rock in the photo above has been eroded by the action of wind

Questions

1 Explain in your own words the following terms:
weathering; transport; deposition; erosion.

2 Certain weathering processes are described as chemical, others as mechanical.
(a) Give two examples of chemical weathering.
(b) Give two examples of mechanical weathering.
(c) Why is mechanical weathering more important than chemical weathering in hot deserts?

3 Look closely at unit G10.
(a) Explain why rain water wears away limestone, writing equations for any reactions involved.
(b) Explain why there are dry valleys in limestone areas.
(c) Explain how limestone pavements form.

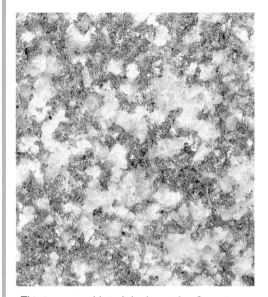

This is a smoothly polished sample of granite—an igneous rock. 'Igneous' means formed by fire. Igneous rocks are formed from the cooling of very hot, molten rock. Granite forms when molten rocks cool slowly below the Earth's surface so the crystals in it are large

Rocks in the Earth are usually mixtures of different substances. When the Earth first cooled, its molten crust solidified to form **igneous rocks**. Initially, there were no other types of rock. Over millions of years, two other types of rock were created—**sedimentary rocks** and **metamorphic rocks**. Figure 1 shows how these three types of rock are being formed today.

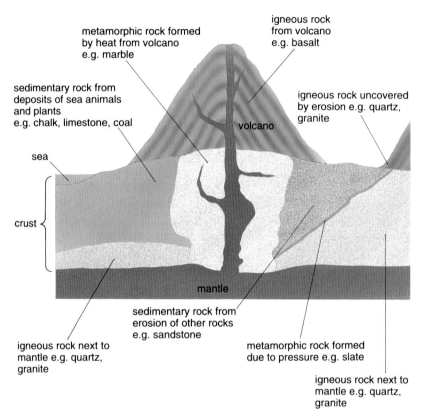

metamorphic rock formed
by heat from volcano
e.g. marble

igneous rock
from volcano
e.g. basalt

sedimentary rock from
deposits of sea animals
and plants
e.g. chalk, limestone, coal

igneous rock uncovered
by erosion e.g. quartz,
granite

sea

volcano

crust

mantle

sedimentary rock from
erosion of other rocks
e.g. sandstone

igneous rock next to
mantle e.g. quartz,
granite

metamorphic rock formed
due to pressure e.g. slate

igneous rock next to
mantle e.g. quartz,
granite

Figure 1
The formation of igneous, sedimentary and metamorphic rocks

Igneous rocks

Igneous rocks have formed from the molten rock or **magma** in the Earth's mantle. The size of crystals in the igneous rock depends on the rate at which the magma has cooled and crystallized. Some igneous rocks are produced when volcanoes erupt and the lava cools quickly in a matter of days or weeks. This produces rocks, such as basalt, with small crystals. Other igneous rocks are formed deep in the Earth's crust next to the mantle. Here the magma cools very slowly, possibly over centuries. This produces rocks with much larger crystals such as granite and quartz.

This is a polished sample of red sandstone—a sedimentary rock. It has formed from the erosion of granite

Sedimentary rocks

When igneous rocks are weathered, they form sediments such as sand and gravel. These sediments may be carried by rivers or ocean currents and deposited elsewhere. As the layers of sediment build up over millions of years, the material below is compressed forming soft rocks such as coal, sandstone and chalk. If the layers are buried deeper, the soft sediments get converted to harder sedimentary rocks like limestone. All of these rocks are known as sedimentary rocks because they are formed by the build up of sediments.

Some sedimentary rocks, like sandstone, result from the weathering of other rocks by wind and water. Other sedimentary rocks, such as chalk, limestone and coal, have formed from the remains of dead animals and plants.

Metamorphic rocks

Sometimes, sedimentary rocks can be changed into harder rocks by enormous pressure or very high temperatures. The new rock has a different structure from the original rock. It is therefore called 'metamorphic rock' from a Greek word meaning 'change of shape'. Slate and marble are good examples of metamorphic rocks. Slate is formed when clay and mud are subjected to very high temperatures. Marble is formed when limestone comes into contact with hot igneous rock.

Deep inside the Earth, the temperature gets so high that the solid rocks melt to form magma. This will eventually solidify as igneous rock beginning the rock cycle again.

Figure 2 shows the stages of the rock cycle. The complete cycle lasts hundreds of millions of years. Notice that there are some short cuts to the full cycle. For example, sedimentary and metamorphic rock can also weather and erode to form sediment. Extreme heat and pressure can also change igneous rocks into metamorphic rocks.

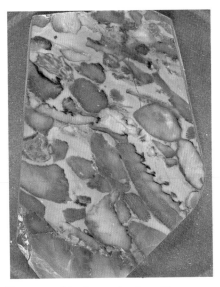

This is a polished sample of marble a metamorphic rock. It has formed from sedimentary rocks at high temperature and pressure

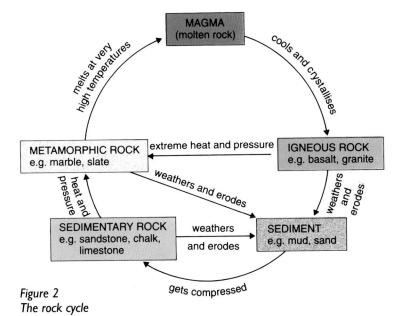

Figure 2
The rock cycle

Questions

1 Look closely at the rock cycle in figure 2 and explain the following statement.
'One rock is the raw material for another'.

2 (a) When the Earth was first formed, there were only igneous rocks. Why was this?
(b) Are the following rocks igneous, sedimentary or metamorphic: (i) coal, (ii) diamond, (iii) slate, (iv) mudstone?

3 (a) Describe how sediments may form on land and in the sea.
(b) How are they turned into sedimentary rocks?

4 You are given three pieces of rock by a friend who wants to know more about them.
Rock A is made of rounded pebbles held together by a hard, sandy layer.
Rock B is white in colour and contains several small fossils which look like shells you have seen at the seaside.
Rock C is very hard and is made of large crystals which you can see quite clearly.
What could you tell your friends about the rocks?

The Earth is shaped like an orange, spherical but slightly flattened at the poles. Its structure is like a badly cracked egg. The 'cracked shell' is the thin crust, the 'white' is the mantle and the 'yolk' is the core (figure 1).

Notice in figure 1 that the three concentric layers of crust, mantle and core increase in thickness, in density and in temperature towards the centre.

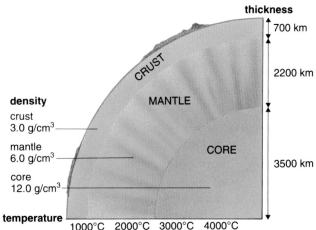

Figure 1

A cross section of the Earth showing the thicknesses, densities and temperatures in its internal structure

Evidence for the Earth's structure comes from various sources. Information about the crust comes from studying rocks, mines, volcanoes and earthquakes. Evidence for the mantle and the core comes from the study of volcanoes, earthquakes and the Earth's magnetic field.

The Earth's crust is cracked and broken into huge sections called **plates**. These vast plates float on the denser mantle below. Over long periods of geological time, the plates move very slowly due to convection currents in the liquid mantle. When the plates slide past each other, move apart or push towards each other, various things can happen.

● **When two plates slide past each other**, stresses and strains build up in the Earth's crust. This may cause the plates to bend. In some cases, the stresses and strains are released suddenly. The Earth moves, the ground shakes violently in an **earthquake** and breaks appear in the ground. These breaks in the ground are called **faults** (figure 2(b)).

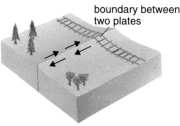

Figure 2
How an earthquake occurs

(a) Plates in the Earth's crust are bent as they slide past each other in opposite directions. The earth and rocks are displaced sideways.

(b) Stresses in the bent planes are suddenly released as a break appears in the earth. The ground shakes (an earthquake) and a fault has formed.

● **When plates move apart**, the crust is being stretched under tension and cracks appear in the Earth's surface. As the plates move further apart, surface rocks can sink forming vertical faults. When two vertical faults occur alongside each other, rift valleys are formed (figure 3).

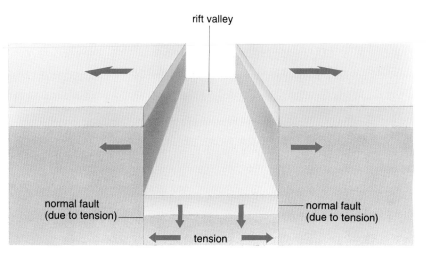

Figure 3
As plates move apart, the land on one side may sink into the crack. If there are two vertical faults parallel to each other, a rift valley may form

In some cases, hot, molten rock escapes through the cracks and erupts as a volcano (unit 3). This can result in ocean ridges.

● **When plates push towards each other and collide**, rocks are squeezed together. When this happens, layers of the Earth's crust are pushed over each other into folds.

Folds in the outer part of the Earth's crust at Stair Hole near Lulworth in Dorset

Mount St. Helens in North America erupted in 1980

Questions

1 The crust of the Earth is thought to account for only 1% of the total volume of the Earth. The mantle accounts for about 82%. Using these figures, construct a pie chart for the major layers of the Earth.

2 Suppose the photo of the folds in the Earth's crust at Stair Hole, Lulworth is taken facing west.
  (a) Which directions did the forces which created the fold come from?
  (b) Was the force a compression or a tension?
  (c) Draw a sketch map of the fold and explain how it formed.

3 In California, most of the orange groves have trees growing in straight lines. In some groves, however, the lines of trees are kinked, although they were not planted like this.
  (a) Explain why the lines of trees are now kinked.
  (b) In what interesting geological area are these orange groves?

4 Although some very deep holes have been drilled into the Earth, none has ever reached the mantle. We have no direct evidence of the nature of the interior of the Earth. Our knowledge is based on indirect evidence. What is this evidence and what has it told us about the Earth's interior?

5 Plate Tectonics

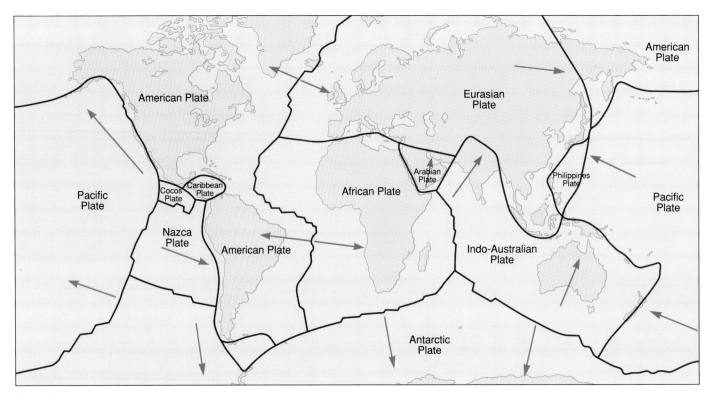

Figure 1
World map showing the main plates and their directions of movement

The study of the movement and interaction of the giant plates on the Earth' surface is called **plate tectonics**. Plate tectonics explains many of the large-scale geological features on the Earth.

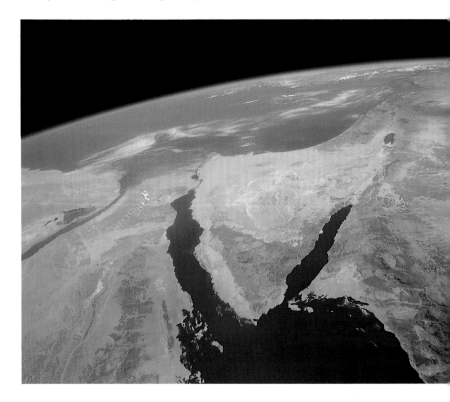

Many years ago, Africa and Arabia used to be joined together. They are now growing apart as the Red Sea gradually widens. This sea may eventually become as large as the Atlantic Ocean

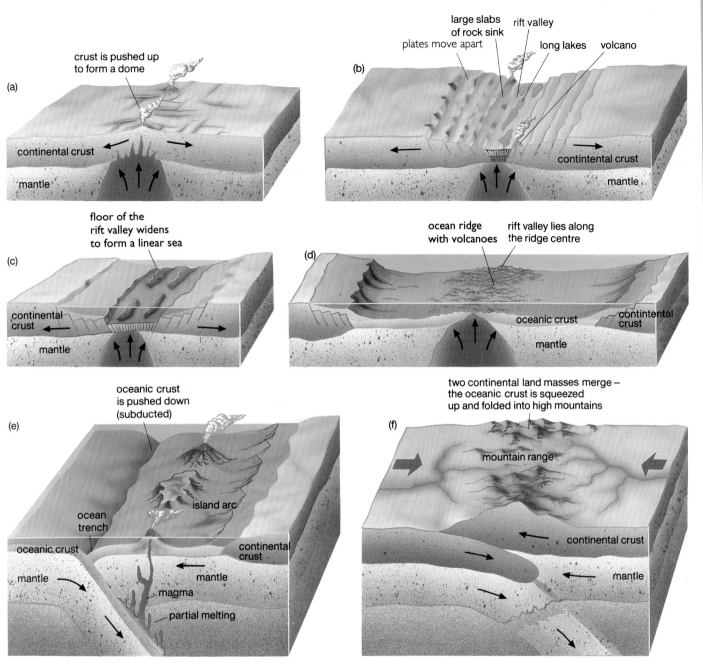

(a) crust is pushed up to form a dome

continental crust

mantle

(b) large slabs of rock sink
plates move apart
rift valley
long lakes
volcano

contintental crust

mantle

floor of the rift valley widens to form a linear sea

(c) continental crust

mantle

(d) ocean ridge with volcanoes
rift valley lies along the ridge centre

oceanic crust

contintental crust

mantle

oceanic crust is pushed down (subducted)

(e) ocean trench
island arc

oceanic crust
mantle
magma
partial melting
continental crust
mantle

two continental land masses merge – the oceanic crust is squeezed up and folded into high mountains

(f) mountain range

continental crust

mantle

The Earth's core is as hot as the surface of the Sun. This causes slow convection currents in the liquid mantle which result in slow movements in the plates of the Earth's crust. Figure 1 on page 274 shows how these plates are moving.

The convection currents circulating in the liquid mantle may take millions of years to rise to the surface. If currents of hot, molten rock rise in places where the Earth's crust is thin, they form **'hot spots'** of intense volcanic activity. The Hawaiian islands lie above one of these hot spots. On these islands, magma rises to the crust surface forming **volcanoes**. In areas where the crust is thicker, rising convection currents push the crust up into a dome, causing tension and cracking (figure 2(a)). If this causes plates on the crust to move

Figure 2
(a) The continental crust is pushed up into a dome
(b) When plates move apart, rift valleys form with volcanoes
(c) A linear sea develops
(d) An ocean ridge forms as the ocean widens
(e) As more sediment collects, the oceanic crust is subducted (pushed down)
(f) As plates move towards each other and continents collide, the ocean closes and a larger land mass with high mountains is formed

These mountains were formed from horizontal beds of sediment at the bottom of an ancient ocean. As the continents moved together and the ocean closed up, this sediment was pushed up into a fold mountain range

apart, large slabs of rock sink and rift valleys form. Volcanoes appear where the magma escapes from cracks in the rift valley (figure 2(b)). This is typical of the rift valleys in East Africa.

The Red Sea and Gulf of Aden illustrate the next stage of the **tectonic cycle**. If the floor of the rift valley widens and deepens, a **linear sea** forms (figure 2(c)). If the African and Arabian plates continue to move apart, this region will gradually form a new ocean. The Atlantic is an example of an expanding **young ocean**. It is widening as fast as your fingernails grow. A **mid-ocean ridge** punctuated with volcanoes has formed down the centre of the Atlantic from the Arctic to the Antarctic (figure 2(d)). This submerged ridge has resulted from the vast amounts of lava pouring out from volcanoes along its length. It follows the boundary between the American Plate and the African and Eurasian Plates (figure 1).

In some places, the volcanic activity is so great that the ocean ridge rises above the surface of the ocean as volcanic islands. Islands of this nature extend all the way down the western side of the Pacific from the Aleutian Islands in the north, through Japan to Tonga in the south.

Over millions and millions of years, thick deposits of sediment collect on the ocean floor. The layers of sediment are particularly thick near continents where rivers have carried silt into the ocean. As the oceanic crust is thin, the weight of sediment makes it sag. More sediment collects, and eventually the crust breaks. As new crust is being created at the ocean ridge, the old oceanic crust, near to the continent, is pushed down or **subducted** (figure 2(e)). The deep depression in the crust where subduction occurs is called an **ocean trench**.

As the Atlantic grows, the American continents are moving westwards. Along the western edge of South America, the Nazca Plate is moving towards the American Plate (figure 1). Here the ocean crust is being forced under the advancing landmass. This is pushing up the continental crust and creating the Andes mountains.

A tectonic cycle ends when two continental land masses converge and the ocean between them disappears. The layers of oceanic crust are squeezed into tight folds forming high mountains (figure 2(f)). This is what happened when India moved north to collide with Asia. The ancient Tethys Ocean disappeared and the Himalayan mountains were formed. The crust here is so thick that volcanic activity has effectively stopped, although earthquakes are common.

Plenty of geological evidence supports the theory of plate tectonics. The latest evidence comes from lasers mounted on satellites which can measure oceanic growth of only 2–3 centimetres a year.

If the Earth was filmed from outer space using time lapse photography over a period of 1000 million years, you would be able to see the **tectonic cycle** clearly: continents moving apart, oceans waxing and waning, new continents joining and mountain chains forming.

Questions

1 Using an atlas, find named examples of volcanoes for each of the following stages in the tectonic cycle (the areas in brackets will help you):

(a) continental rifting (East African Rift valley),

(b) linear sea spreading (Red Sea and Gulf of Aden),

(c) growth of ocean (islands on the Atlantic ridge),

(d) subduction of ocean floor near to continents (the Andes and the Rockies).

Are there any volcanoes in the final stage of the cycle, (e.g. the Himalayas)? Are there any other locations where volcanoes can be found? If so, why are they found there?

2 Draw up a flow chart to show the stages of the tectonic cycle.

Section L: Activities

1 Rock on the Map

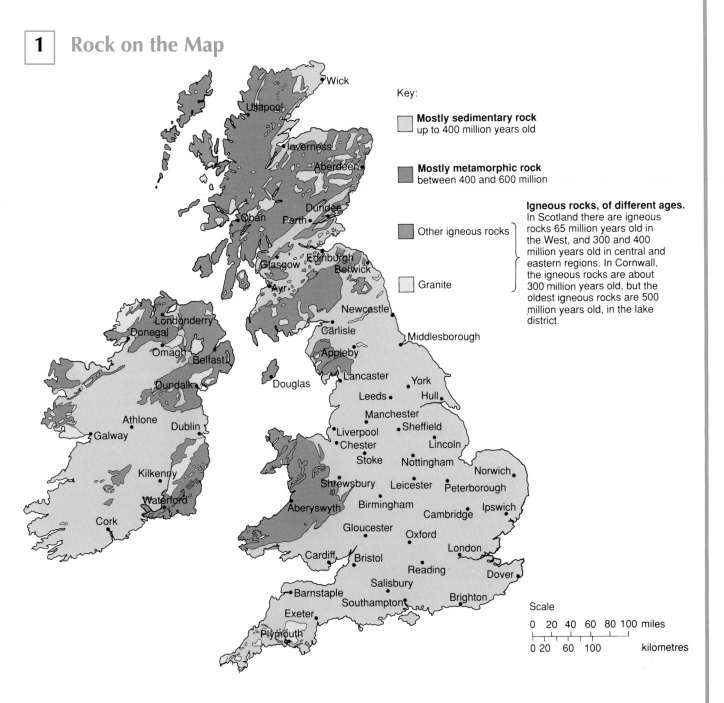

Key:

☐ **Mostly sedimentary rock**
up to 400 million years old

■ **Mostly metamorphic rock**
between 400 and 600 million

■ Other igneous rocks ⎱

☐ Granite ⎰

Igneous rocks, of different ages.
In Scotland there are igneous rocks 65 million years old in the West, and 300 and 400 million years old in central and eastern regions. In Cornwall, the igneous rocks are about 300 million years old, but the oldest igneous rocks are 500 million years old, in the lake district.

Scale

0 20 40 60 80 100 miles

0 20   60   100   kilometres

This map shows the type of rock that lie at or just beneath the surface in the British Isles. Using the map, the key and perhaps an atlas, answer the following questions.

1 About how old is the rock upon which these places are built?
  a) Liverpool   b) Aberystwyth
  c) Aberdeen   d) your home
2 What type of rock are these places built upon?
  a) Plymouth   b) Oban
  c) Belfast   d) your home

3 How far away are the nearest igneous rocks to
  a) London   b) Edinburgh
  c) Newcastle   d) your home?
4 How long ago were the most recent volcanoes active in the British Isles
5 About 400 million years ago, part of the British Isles was a huge mountain range, similar to the Himalayas today. This mountain range contained metamorphic rocks and granite. From the map, which part of the British Isles do you think this was?

2 The table shows data obtained by school children doing a weathering project in a graveyard.

Dates on gravestones	Type of rock		
	Sandstone	Marble	Igneous
1720–1770	extremely badly weathered	very badly weathered	moderately weathered
1771–1819	very badly weathered	badly weathered	moderately weathered
1820–1870	badly weathered	moderately weathered	slightly weathered
1871–1919	moderately weathered	slightly weathered	unweathered
1920–1970	slightly weathered	unweathered	unweathered
1971–1991	unweathered	unweathered	unweathered

(a) Why do you think they chose a graveyard for this study?
(b) What do the results show?

3 The graph below shows the velocity of flow of a stream of water needed to pick up and *transport* particles of different sizes (Curve A). Curve B of the graph shows the velocity of flow at which particles of different sizes come out of suspension and are *deposited*. Use the graph to answer the following questions.

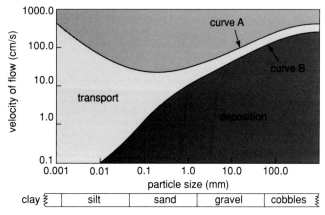

(a) Which type of material is most easily transported?
(b) At approximately what velocity will clay particles be picked up?
(c) At approximately what velocity will coarse gravel be picked up?

(d) In a stream flowing at a velocity below 50 cm per second, what type of material is most likely to be picked up?
(e) Explain the shape of Curve B.

4 The map below shows the location of three seismological stations which record earthquakes. The time that each station receives information on its equipment depends upon its distance from the epicentre of the earthquake. The table shows how far each station is from the epicentre.

Continent	Distance of station from epicentre/km	
	Quake 1	Quake 2
A	2500	1800
B	2000	1500
C	1100	3200

(a) Trace the map accurately and, using a compass, construct arcs to locate each epicentre.
(b) Why is it essential to know three distances?
(c) Why do stations receive little data for some earthquakes?

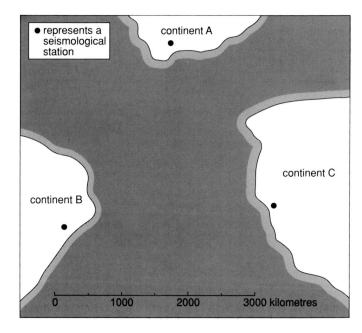

Section L: Study Questions

1 What are the main sources of carbon dioxide in the atmosphere? What is the role of carbon dioxide in the atmosphere? What could happen if atmospheric levels of carbon dioxide changed significantly?

2 What is the significance of ozone in the atmosphere? What is causing the amount of ozone to change? What may be the long term consequences of increasing or decreasing the amount of ozone in the atmosphere?

3 The table below shows the major *natural* reservoirs for water on the Earth. The volume of water in each of these reservoirs is given in cubic metres.

Natural reservoir	Volume of water held in reservoir/m^3
Oceans	1350×10^{15} m^3
Glaciers and polar ice	29×10^{15} m^3
Underground water in rocks and natural wells	8.4×10^{15} m^3
Lakes and rivers	0.2×10^{15} m^3
The atmosphere	0.013×10^{15} m^3
The biosphere	0.0006×10^{15} m^3

(a) Calculate the amounts of water held by the oceans as a percentage of the total water in natural reservoirs.
(b) What percentage of the total in natural reservoirs is found as glaciers and polar ice?
(c) How much water is held in the atmosphere as a percentage of all *non-ocean* water?

4 An **aquifier** is a rock which holds water. An **aquiclude** is a rock which will not hold water and through which water cannot pass. An **artesian basin** is a region where water is trapped in an aquifer, often far below the surface. Using the geological cross-section of the London Basin, explain why London has such a good water supply.

5 The diagram below shows an alpine area. Make a simple copy of the diagram. By each label line, put the appropriate letter(s) as follows: **'W'** for weathering, **'E'** for erosion, **'T'** for transport, **'D'** for deposition. If you know the names of the features shown, you can add these to your diagram too.

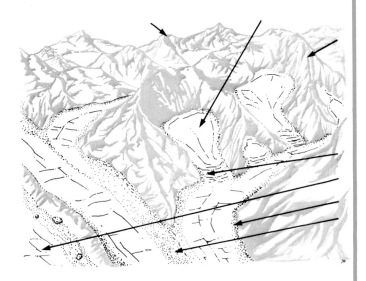

6 The following diagram shows three samples of rock material. They are the same rock type and have an equal volume. Calculate the total surface area for each sample. Which sample would be most susceptible to weathering processes? How will their rates of weathering compare?

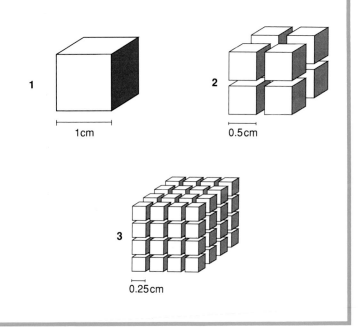

Index